a practical guide to

ELISA

Pergamon titles of related interest

a practical guide to

ELISA

by
D M KEMENY

Department of Allergy
Guy's Hospital, London

PERGAMON PRESS
Member of Maxwell Macmillan Pergamon Publishing Corporation
OXFORD · NEW YORK · BEIJING · FRANKFURT
SÃO PAULO · SYDNEY · TOKYO · TORONTO

U.K.	Pergamon Press plc, Headington Hill Hall, Oxford OX3 0BW, England
U.S.A.	Pergamon Press, Inc., Maxwell House, Fairview Park, Elmsford, New York 10523, U.S.A.
PEOPLE'S REPUBLIC OF CHINA	Pergamon Press, Room 4037, Qianmen Hotel, Beijing, People's Republic of China
FEDERAL REPUBLIC OF GERMANY	Pergamon Press GmbH, Hammerweg 6, D-6242 Kronberg, Federal Republic of Germany
BRAZIL	Pergamon Editora Ltda, Rua Eça de Queiros, 346, CEP 04011, Paraiso, São Paulo, Brazil
AUSTRALIA	Pergamon Press Australia Pty Ltd., P.O. Box 544, Potts Point, N.S.W. 2011, Austrailia
JAPAN	Pergamon Press, 5th Floor, Matsuoka Central Building, 1-7-1 Nishishinjuku, Shinjuku-ku, Tokyo 160, Japan
CANADA	Pergamon Press Canada Ltd., Suite No. 271, 253 College Street, Toronto, Ontario, Canada M5T 1R5

First edition 1991

Library of Congress Cataloging-in-Publication Data
Kemeny, D. M.
A practical guide to Elisa / by D. M. Kemeny.
p. cm.
Includes bibliographical references
1. Enzyme–linked immunosorbent assay. I. Title.
QP519.9.E48K46 1990 616.07'9—dc20 90-46231

British Library Cataloguing in Publication Data
Kemeny, D.M.
A practical guide to ELISA.
1. Medicine. Immunoassay. Use of enzymes. Techniques
I. Title
616.0756
ISBN 0-08-037508-1   Hardcover
ISBN 0-08-037507-3   Flexicover

Printed in Great Britain by B.P.C.C. Wheatons Ltd., Exeter

Contents

Contents

Contents

Contents

Acknowledgements

I would like to thank the following: my wife Hilary and daughter Louise who put up with me spending so many hours closeted with the computer; Dave Richards for providing such common sense as well as imagination to our experiments; Mac Turner, Hilary Kemeny, Marion Jowett, Val Corrigal for their helpful comments and criticism; Mr Ted Hill and Tom and Brenda Cowan in whose cottages I was able to write this book; the inventors of the Apple Mac computer; and finally all those whose people whose problems with ELISA encouraged me to come to grips with this technique. There is plenty to challenge the intellect in this apparently simple method.

Preface

The aim of this book is to provide a guide to ELISA, both for those who are new to the technique and for those who already have some experience of it. I would like to provide the reader with a systematic approach to setting up these assays and to warn him or her of some of the pitfalls. The first part of the book deals with rudimentary aspects of human immunology and the principles underlying immunoassays. The second part describes the various forms that ELISA takes and offers an approach to setting up assays. In addition, this section details the preparation of reagents for ELISA and considers how to get the best out of them. Finally, there is an appendix which contains practical details of the various techniques that are used in my laboratory. This is not intended to be comprehensive - you cannot advise about methods you have not used yourself, but it is hoped that this will provide some useful information. I would also like to direct the reader to the books listed in the bibliography which I believe are a valuable additional source of information. In an attempt to make the book more readable, detail of methods and references have been restricted to a brief bibliography. As far as possible I have tried to avoid the use of specialist terms or jargon. As a friend, herself a scientist said, on being asked what she thought about immunology - "It's all alphabet soup to me!".

D. M. Kemeny

CHAPTER 1

Immune Cells
and Antibodies

The immune system is the body's mechanism for fighting disease. Infection can be prevented by simple physical barriers, such as the skin, but the body is vulnerable at exposed sites (mouth, nose, lung, eyes, gut, urinary and genital tracts). Once infectious agents have penetrated its interior, a specialised set of cells and their products are brought into play.

These include lymphocytes, macrophages and granulocytes and have evolved to combat infection. They either exert direct effects on invading organisms (T cells and macrophages) or produce specialised molecules called antibodies which attack the invading organism (B cells). There are, in addition, regulatory, inflammatory and accessory cells which regulate and amplify the immune reaction.

LYMPHOCYTES

B cells

These are the cells that make antibodies. They are mainly found in specialised tissues: lymph nodes, spleen. Once stimulated, they divide rapidly to generate sufficient B cells capable of producing enough of a particular antibody to fight the infection. An activated B cell is called a plasma cell (Fig 1.1). Each plasma cell secretes one type (class) of antibody and all the antibody produced by a single cell is of the same specificity. The methods used for studying antibody production from cultured B cells using ELISA will be discussed in chapter 3.

T cells

T cells fulfil two major functions. They regulate the activity of B cells and directly attack infectious organisms. Regulatory T cells can both support B cell division and also determine the class of antibody produced. In some circumstances T cells can damp down or suppress antibody responses, for example when an infection is over.

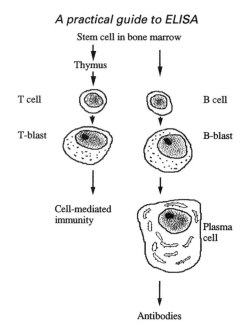

FIG 1.1 Lymphocytes

In order to carry out many of these activities T cells secrete a number of specialised molecules called lymphokines. The methods for studying lymphokine production by ELISA will be discussed in chapter 3.

ANTIBODIES

All antibodies belong to a special category of protein called immunoglobulin, usually abbreviated to Ig. The function of these specialised molecules is to bind to foreign proteins (antigens) and activate specific reactions (cell killing, inflammation, prevention of bacterial and viral attachment). They can be divided into two parts which provide specificity and effector function (Fig 1.2). At one end there is a combining site which binds to antigen and at the other are specific receptors for a number of other molecules such a complement (which can punch holes in cells) and secretory component

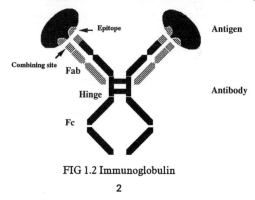

FIG 1.2 Immunoglobulin

(which prevents degradation by digestive enzymes), and determinants bound by special cell receptors (such as IgE receptors on mast cells). They can be divided into five classes (Table 1.1) according to their specific effector functions.

TABLE 1.1.
The different classes of human immunoglobulin.

Class	Heavy chain	Serum conc	Molecular weight (mg/ml)	Form	Crosses placenta	J chain	Subclasses	Serum half-life (days)	Complement fixation
IgA	α	0.5-3	160,000	monomer	no	no	1,2	21	no
			400,000	dimer	no	yes			
IgD	δ	0.003	180,000	monomer	no	no	none	21	no
IgE	ε	0.0001	190,000	monomer	no	no	none	2.5	no
IgG	γ	10-13	150,000	monomer	yes	no	1	21	yes
							2, 4	21	no
							3	7	yes
IgM	μ	0.5-2.5	900,000	pentamer	no	yes	none	?	yes

IgA

Two forms of IgA exist, serum and secretory IgA. The latter is found in the mucous secretions and is protected from digestion by a molecule called secretory component. Secretory IgA is always found as a dimer. The heavy chains are attached by a joining chain (J chain) and can be regarded functionally as 'antiseptic paint' providing a first line of defence which prevents infectious organisms from entering the body. It is probably most important in early life where it may limit the extent of infection and so 'buy time' for the rest of the maturing immune system.

IgD

This is found on the surface of immature B cells along with IgM. A role for this class of immunoglobulin has yet to be determined.

IgE

This is the most recently discovered class of immunoglobulin. It is found at the lowest serum concentration of all the immunoglobulins and is responsible for immediate-type allergic conditions such as hay fever, asthma, food, drug and insect sting allergy. It causes these reactions by activating specialised inflammatory cells (mast cells and basophils) to which it binds through specific receptors (Fig 1.3). These receptors (FcERI) have very high affinity (strong binding) for IgE. In addition IgE binds to lower affinity receptors (FcERII) on macrophages, eosinophils and platelets.

IgG

Quantitatively this is the major class of immunoglobulin in serum. It has

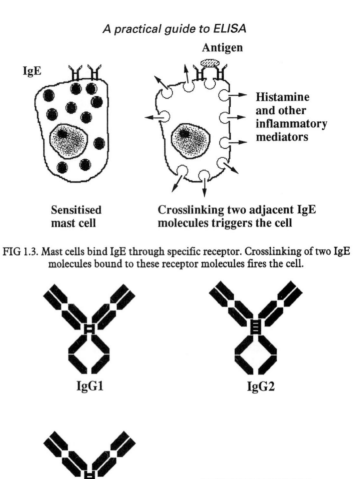

FIG 1.3. Mast cells bind IgE through specific receptor. Crosslinking of two IgE molecules bound to these receptor molecules fires the cell.

FIG 1.4. The four subclasses of human IgG showing the differences in hinge region.

four different forms or subclasses, two of which can bind to and activate the complement system (a family of molecules which are inflammatory and have the capacity to form holes in cells so causing them to burst and die). All four subclasses of IgG possess receptors for the first complement component, C1q. The reason why IgG1 and IgG3 fix complement but IgG2 and IgG4 do not probably resides in the flexibility of the hinge region (Fig 1.4) which determines access to the C1q receptor. All four subclasses of IgG are transported from mother to fetus and provide newborns with the full array of maternal antibodies for the first 3 months of life, until they can make their own.

IgM

IgM antibodies are produced first in an infection but are only present for a relatively short time and are of low affinity. Like IgA they can bind a short length of a connecting protein called 'J-chain' as well as to secretory piece. While IgA forms dimers, IgM forms pentamers which endow it with the greatest avidity for complement of all the immunoglobulins. This makes it the most potent immunoglobulin at activating complement and so especially useful in fighting infection.

Affinity

The specific binding of antibody to antigen occurs at the combining site. Here complementary structures on the antibody (the combining site) and the antigen (the antigenic determinant or epitope) form the link between the two molecules. The strength of binding between them is called affinity and is the product of the various chemical forces that attract the antibody combining site and its corresponding epitope. Not only should the appropriate charges be on the opposite sides of the combining site but the shape of the epitope and that of the antibody binding site should permit close contact between the two reactive sites. This is termed the goodness of fit.

ANTIBODY	Fab	IgG	IgG	IgM
effective antibody valence	1	1	2	up to 10
antigen valence	1	1	n	n
equilibrium constant	10^4	10^4	10^7	10^{11}
advantage of multivalence	–	–	10^3 - fold	10^7 - fold
definition of binding	affinity	affinity	avidity	avidity
	intrinsic affinity		functional affinity	

FIG 1.5. Affinity and avidity of antibody (published with the kind permission of Dr ME Devey and Dr MW Steward).

Avidity

Some antibodies of low affinity can still bind strongly if they do so with more than one epitope at the same time. This is referred to as the valency or avidity of the interaction. Just as a gambler's odds are multiplied together for naming both the first and second horses in a race, so the probability that both antibody combining sites would let go of their respective epitopes at the same time is less than for a single interaction (Fig 1.5). The effect of multiple attachment between antibody and antigen is to combine the respective affinities. This combined affinity is called functional affinity.

CHAPTER 2

Immunoassays

An immunoassay is a technique for measuring the presence of a substance using an immunological reaction. While this could be used to describe the proliferation of lymphocytes in culture, or the wheal and flare response of the skin of an allergic individual when allergen is injected, it is almost exclusively used to describe tests that exploit the reaction between antibody and antigen in vitro. Immunoassays include simple precipitation of antibody/antigen complexes, agglutination of coated red cells or other particles, electrophoretic immune precipitation in agar, and radio (RIA) and enzyme immunoassays (EIA). In this chapter I shall describe the spectrum of immunoassays and discuss the reasons for choosing one over another.

A number of factors determine the choice of assay: sample concentration, precision, ease of use, safety and available laboratory facilities. Perhaps the most important of these is sample concentration. This determines the scale of the assay. Just as a carpenter would not use a pin hammer to drive in 6-inch nails or a 14 lb sledge hammer for cabinet making, so different types of assay are suited to measuring different amounts of a particular material. A student I taught recently told me of the difficulties she was having with an enzyme-linked immunosorbent assay (ELISA) for serum IgG, A and M. It had been suggested by colleagues that single radial immunodiffusion would have been more suitable and they were, of course, quite right; ELISA was far too sensitive. It is essential, therefore, that you consider how much sample you are measuring.

In historical terms, the more sensitive RIA and EIA techniques were developed more recently than less sensitive precipitation and agglutination based assays. However, just because some of these techniques were discovered a long time ago it doesn't mean that they aren't still useful. As I describe the various assays below I have attempted to indicate their optimal working range and the circumstances in which they might be most appropriate.

PRECIPITATION ASSAYS

When an antibody combines with an antigen it forms an immune complex. In antigen excess all the antibody binding sites are occupied, resulting in the formation of small, soluble complexes of antibody/antigen (Fig 2.1a). If the antibody is in excess all the binding sites on the antigen may be saturated and again small soluble complexes are formed (Fig 2.1b). But if neither is is excess

7

then some of the reactions between antibody and antigen will be with adjacent molecules resulting in the formation of an antibody/antigen polymer or lattice (Fig 2.1c). Together with bound complement this can involve a great many molecules. This large immune complex is poorly soluble and so precipitates. Precipitation occurs when the ratio of antibody to antigen favours multiple attachments between them, and this point is termed equivalence (Fig 2.2).

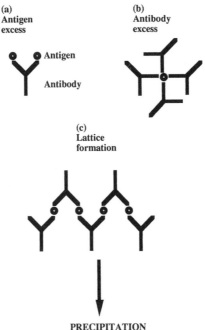

FIG 2.1.Immune precipitation

Gel diffusion assays

When a solution of antigen is placed in a recess or well cut in an agar-coated glass slide the proteins will diffuse away from the well into the agar in all directions, in much the same way as a drop of ink spreads on a piece of blotting paper. If the appropriate antibody is present in the agar then a precipitate will form at equivalence. This principle is exploited in gel diffusion tests.

Double diffusion

Double diffusion is one of the most technically simple forms of immune test. It involves putting a drop of antigen in a well cut in an agar-coated glass plate and antibody in an adjacent well. The two diffuse outwards and a precipitate forms between the two wells (Fig 2.3). Precipitation occurs at equivalence (see Fig 2.2). The precipitate often appears as a line which is called a precipitin line. The method can usefully detect between 0.1 and 10 mg/ml of antibody or antigen, it is easy to perform and is ideal for monitoring antibody

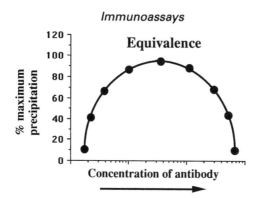

FIG 2.2. As antibody concentration is increased so the amount of precipitate increases until equivalence is reached. In antibody excess the amount of precipitate declines.

(a)

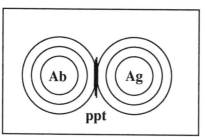

(b)

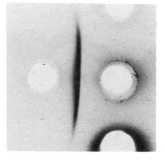

FIG 2.3. (a) The principle of double gel diffusion. Antibody (Ab) and antigen (Ag) diffuse away from the wells cut in the agar coated glass slide. They diffuse outwards until they meet and precipitate (ppt). (b) A photograph of a precipitin line between IgE and anti-IgE.

responses in immunised animals. No specialised equipment is needed and reagents can usually be used without processing. It can also be used to compare the specificity of different antibodies. If they are of the same specificity, adjacent precipitin lines fuse, if different then they will cross over. It is possible to amplify the precipitate formed with additional antibodies or to use sensitive stains to increase sensitivity down to 0.1 µg/ml. Addition of polyethylene glycol (PEG) to the agar can be used to facilitate precipitation.

Single radial immunodiffusion

While some idea of quantity of antibody or antigen can be gained from the position of the precipitate formed in double diffusion, the precipitin line will be furthest from the reagent present at the highest concentration. It is easier, however, to obtain quantitative data if the test antigen diffuses into agar jelly containing the relevant antiserum. It will do so until it reaches the point of equivalence. The diameter of the ring of precipitate (Fig 2.4) is directly proportional to the concentration of antigen. The amount of sample present can be determined by reference to a calibration curve using known standards (see chapter 7).

Electrophoresis

All proteins carry electrical charges. At a given pH these will result in a net positive or negative charge. In a suitable medium, proteins can be transported in an electric current to an electrode of the opposite charge. This process is called electrophoresis and can be used to increase the sensitivity and the discrimination of the immune precipitation reaction.

Immunoelectrophoresis

It is possible to analyse antigens present in complex mixtures of proteins, such as are found in serum, or in bacterial or allergen extracts, by a

(a)

Ab in the gel

ppt   Ag

(b)

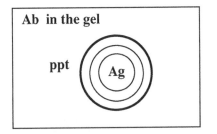

FIG 2.4. (a) The principle of single radial immunodiffusion. Antigen (Ag) is placed in a well cut in an agar-coated plate. The agar contains the appropriate antibody (Ab). At equivalence the diffusing antigen precipitates. The larger the ring the greater the concentration of Ag. (b) A photograph of precipitin rings from an IgG subclass assay.

combination of electrophoresis and gel diffusion. The antigens are first separated electrophoretically and then a trough is cut in the agar adjacent to the run of the antigens and antibody added to it. After an 18 hour incubation, a series of precipitin arcs can be seen (Fig 2.5). These correspond to the different antibody/antigen reactions that have taken place. This assay is called immunoelectrophoresis and and has tremendous resolving power. We use it routinely to test the purity of isolated monoclonal antisera. Albumin, for example, often contaminates such preparations and can be seen at the anodic

(a)

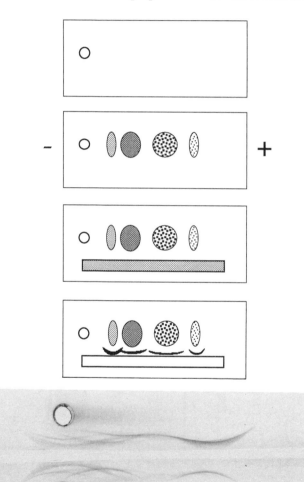

(b)

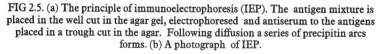

FIG 2.5. (a) The principle of immunoelectrophoresis (IEP). The antigen mixture is placed in the well cut in the agar gel, electrophoresed and antiserum to the antigens placed in a trough cut in the agar. Following diffusion a series of precipitin arcs forms. (b) A photograph of IEP.

end of the gel. Once purified, by protein A or protein G, a single precipitin arc can be seen.

Rocket immunoelectrophoresis

It usually takes 12-18 hours for an immune precipitate to form in agar. If, however, the antigen is electrophoresed into agar jelly containing antibody at pH 8.6 it is possible to form an immune precipitate much faster.

The precipitate formed has the appearance of rockets (Fig 2.6) which give their name to the test and as little as 0.01 mg/ml of antigen can be detected. One might ask why the antibodies in the gel do not migrate out of it. The explanation is that at pH 8.6, which is close to the isoelectric point of rabbit IgG (the pH at which the positive and negative charges on the molecule are

(a)

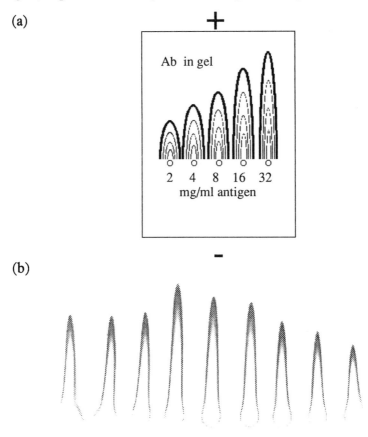

(b)

FIG 2.6. Rocket electrophoresis. (a) Antigen passes from the well into the antibody - containing gel as a result of the electrical current applied. At equivalence the antigen is precipitated and cannot migrate further. (b) An example of rocket electrophoresis (reproduced with the kind permission of Dr A W Ford).

equal), the antibodies carry no charge and so remain static.

Fused rockets

Rocket electrophoresis can be used to analyse proteins separated by a variety of chromatographic procedures. In this version of the method, a series of wells are cut in agar. To these are added samples of the various fractions obtained from the chromatography column, typically one that separates proteins by charge or size. These are allowed to diffuse into the surrounding agar for 45 minutes before being run into a second gel cast next to the first. This gel contains an antiserum which contains antibodies specific for the different proteins in the sample. Thus, separation of the antigens in the sample is performed outside the gel (in the chromatography column) and these subsequently react with antibody incorporated in the gel. Where the same protein is present in two adjacent wells a continuous line of precipitation will be seen (Fig 2.7). As most chromatographic columns elute the sample as a few peaks by optical density, this technique greatly increases the resolving power of these chromatography columns.

Crossed immunoelectrophoresis

The resolving power of the two techniques of immune electrophoresis and rocket electrophoresis can be extended further by running the separated antigens into a gel containing antibodies of a number of specificities which will form precipitin arcs whose position reflects the electrical charge and concentration of the precipitated antigen. The resulting precipitates are better separated and so are easier to see (Fig 2.8).

(a)

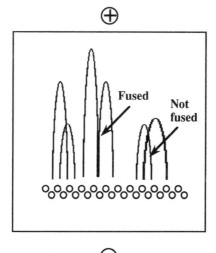

FIG 2.7. (a) The principle of fused rockets.Antigens are separated and placed in the wells and allowed to diffuse. They are then electrophoresed into the main gel containing antibody to the antigen mixture. (Fig 2.7(b) overleaf)

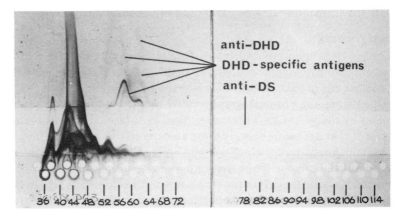

FIG 2.7. (b) Fused rockets of separated dog hair antigens separated by gel filtration
(reproduced with the kind permission of Dr AW Ford).

Radio and enzyme immunoelectrophoresis

The immunological identity of the proteins contained in the precipitates of all these gel techniques can be visualised through the binding of enzyme or radiolabelled antigen or antibody to the immunoprecipitate. With the simple double diffusion test it is possible to add radiolabelled antigen to the well. This will bind to the precipitated antibody and can subsequently be visualised by autoradiography (X ray). Alternatively, human antibody can be incubated with a crossed immunoelectrophoresis gel followed by radiolabelled anti-IgE to identify IgE binding proteins or 'allergens' (Fig 2.9).

(a)

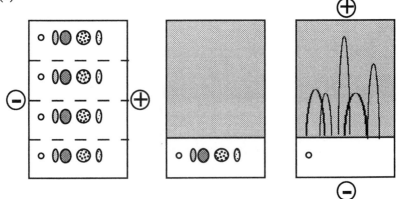

FIG 2.8. (a) The principle of crossed immunoelectrophoresis (CIE). Proteins are separated in the first dimension and then electrophoresed into the second gel containing antibody where they are precipitated.

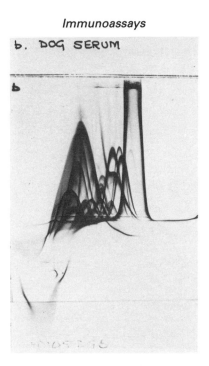

FIG 2.8.(b) A photograph of CIE with dog hair antigens (reproduced with the kind permission of Dr AW Ford).

RADIOIMMUNOASSAY (RIA) AND ENZYME IMMUNOASSAY (EIA)

Techniques for incorporating radioisotopes of high specific activity into proteins were introduced in the late 1950's. Enzyme labels were introduced some ten years later. The sensitivity of immunoassays developed using both radio and enzyme labels were around a 1000 fold greater than that of previous techniques.

Separation versus separation-free assays

Separation-free assays

These exploit a change in enzyme activity that can occur when antibody binds to an antigen labelled with an enzyme. Because the binding of antibody to antigen itself provides the signal, separation of bound and free label is not needed. Alternatively, the separation may take place within the matrix of a solid phase without requiring any wash steps.

Separation assays

Most RIA and EIA techniques require separation of bound and free antibody or antigen, and this book is concerned with this type of assay. Assays

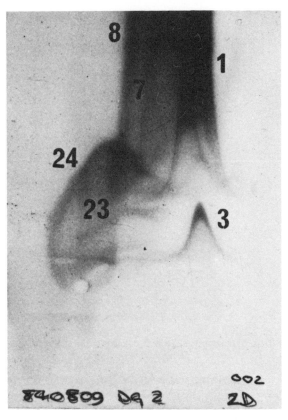

FIG 2.9. Crossed radioimmunoelectrophoresis.

may be carried out in a fluid phase or on a solid phase and sample may be detected in a competitive or a non-competitive fashion.

Competitive versus non-competitive assays

Competitive assays, as their name implies, measure competition in binding to antibody between a fixed amount of labelled antigen and an unknown quantity of unlabelled antigen, 'the sample'. There are many variants and the assay can, of course, be reversed and used to measure the competition of labelled and unlabelled antibody for antigen. Competitive techniques are more   demanding in terms of the accuracy with which the different reagents need to be dispensed and purity of the labelled ligand. They are are easier to quantitate and can be less influenced by contaminants. In non-competitive assays only one component, the sample, is present at a limiting concentration. Thus errors in dispensing other reagents have little or no effect on the result. This type of assay will normally be easier to control and yields accurate results but is more likely to be influenced by cross reactions and non-specific binding.

Competitive assays

In classical RIA, competition between a fixed amount of radiolabelled antigen and an unknown quantity of cold or unlabelled antigen (the test sample) is measured (Fig 2.10). For example, for quantifying IgE in serum, radiolabelled IgE myeloma protein is mixed with the sample containing an unknown amount of IgE and a fixed amount of anti-IgE.

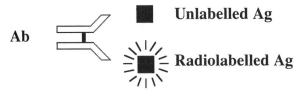

FIG 2.10. Classical RIA. Unlabelled Ag in the sample competes with a fixed amount of radiolabelled Ag for binding to a limited amount of antibody.

The antibody-bound IgE must then be separated from the free IgE by precipitation with a second antibody which can be attached to particles to speed up the process. Other non-specific precipitating agents, such as ammonium sulphate, trichloracetic acid and polyethylene glycol, have also been used. Once bound to protein A-bearing bacteria or protein A- or antibody-coated agarose beads (such as Sepharose) very rapid precipitation (20 seconds in the centrifuge) can be achieved. In competitive RIA the more antigen present in the sample the less radioactive antigen will be bound. These assays can be used to study protein structure, with monoclonal antibodies for example, without being affected by changes in antigen structure due to binding to a solid-phase.

Non-competitive assays

These can involve the simple binding of radiolabelled antigen to antibody and their subsequent precipitation (Fig 2.11). In solid phase, non-competitive assays the antibody or antigen in the sample is first bound to the coated solid phase and is subsequently detected with a second, labelled antibody or antigen

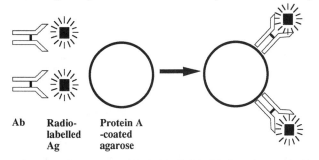

FIG 2.11. Antigen binding assay. Radiolabelled Ag binds to the sample Ab and is precipitated by the protein A-coated agarose particles.

(Fig 2.12). The amount of radio or enzyme activity bound rises in proportion to the concentration of the sample, the exact opposite of competitive assays.

Allergen-coated   IgE Ab in   Radiolabelled
disc   the sample   anti-IgE

FIG 2.12. Immunoradiometric assay (IRMA). A good example of the immunoradiometric assay is the radioallergosorbent test shown above. IgE antibody in the test sample binds to allergen on the paper disc and is subsequently detected with radiolabelled anti-IgE.

Solid versus fluid-phase assays

Another important consideration is the medium in which the assay is carried out. This can be in a solution (fluid phase) or on the surface of plastic or some other protein-binding material (solid-phase). The choice of solid or fluid phase assay depends largely upon what is being measured. There are some inherent advantages in both types of assay but solid-phase techniques are usually much easier to perform.

Fluid-phase assays

The main advantage of fluid-phase assays is that the behaviour of molecules in solution is much easier to predict and the reaction with antibody obeys the the law of mass action:

Association: Antigen and Antibody $\rightarrow$ Antigen-Antibody
                                                    r1   complex

Dissociation: Antigen + Antibody $\leftarrow$ Antigen-Antibody
                                                    r2   complex

Put simply the law of mass action says that the likelihood of antibody of the appropriate specificity coming into contact with the corresponding antigen will depend on the concentration of each in the reaction mixture. The rate at which this happens will depend on their concentration. A good analogy is the unexpected meeting of friends in a large crowd. The larger the crowd, the less likely that such chance encounters will occur.

The likelihood of contact between antibody and antigen is not the only factor that will determine the outcome of such interactions. Once bound, other factors will determine whether they remain so. Imagine, for example, a room full of couples dancing. By the end of the evening it may only be those who have a strong liking for one another that will remain together. In molecular terms the strength of this affection is called affinity. Thus any antibody/antigen interaction will depend on the rates of association (r1) and dissociation (r2).

Solid-phase assays

Solid phase assays are generally easier to perform and more sensitive than fluid phase assays for two reasons. First, it is possible to bind a number of detector molecules to each molecule of the sample. Second, when antibody or antigen dissociates from the surface of the solid-phase, it is likely to remain close to it and so have a greater chance of recombining than if the reaction were taking place in solution (Fig 2.13). The concentration of antibody or antigen at the surface of a solid-phase, in the unstirred layer, is extremely high. Any that diffuses away is likely to remain in close proximity to the solid phase and so be more likely to re-bind. This may explain why these assays can detect antigen or antibody well below the affinities of the reagents involved. The fact that these assays can detect antibodies of lower affinity can, however, sometimes be a drawback as background or non-specific binding tends to be greater.

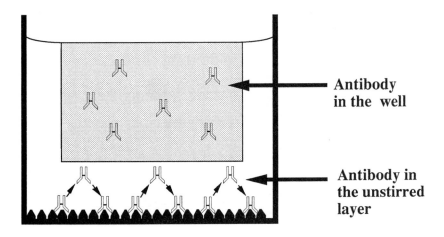

FIG 2.13. Solid-phase assays are less dependent on the affinity of the antibody because there is an unstirred layer of buffer next to its surface. Antibody which becomes dissociated from the antigen on the plate is likely to remain in this layer and so have a much greater chance of recombining than antibody elsewhere in the well which has yet to be bound.

OTHER IMMUNOASSAYS

Nephelometry

Nephelometry is a technique which utilises the change in the light-scattering properties of a solution that occurs following an antibody/antigen interaction. It can be automated, which has resulted in its widespread use in routine laboratories, for example the measurement of immunoglobulin levels.

Agglutination assays

Agglutination is the clumping and sedimentation of particulate antigen after reaction with antibody and was first seen with bacteria after incubation with serum from an infected patient. Agglutination of red blood cells following incubation with serum led to the discovery of ABO blood groups. The reaction depends on antibody bridges made between antigen particles on the red cells by bi- or multi-valent antibody.

Red cell agglutination

Together with gel precipitation techniques this has been in existence for many years. It is sometimes described as haemagglutination. In this technique (Fig. 2.14) red blood cells (typically from sheep) are coated with antibody or antigen. In the absence of the corresponding antibody or antigen the cells will settle at the bottom of a tube or microtitre plate well to form a distinct red button. In the presence of the test sample, the cells are cross-linked and form a "mat". A positive result is a diffuse red colour.

Haemagglutination

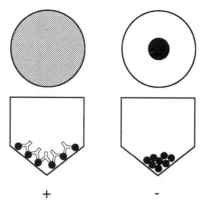

FIG 2.14. Red cell agglutination. In the presence of antibody antigen-coated red cells are linked together and form a diffuse "mat" over the well.

CHAPTER 3

ELISA Design

There are many different ways of configuring ELISAs and the range of assay formats that are available can be bewildering. The choice depends on the nature of the sample, availability of reagents and the precision and sensitivity required. In many situations, it is not necessary to have a precise measure of the concentration of a given sample, for example in the screening of hybridomas or the routine testing of food samples for the presence of additives. Under such circumstances it may be sufficient to record the result as a simple yes or no. Similarly it may be sufficient to record the result as high, medium or low, which is indeed the form in which many routine test results are reported. Such semi-quantitative methods still need to be carefully developed but considerations of speed, sterility, and convenience will also be important. Pin assays are particularly useful here (see chapter 4).

Under exceptional circumstances, it is quite reasonable to use a simple and perhaps less perfect assay. The results will need to be interpreted accordingly but if the answer is clear cut this may be acceptable. More commonly the answers are not so easily obtained and the user must develop assays of high precision and sensitivity. Assays inevitably deteriorate over time and an assay which was initially sensitive and accurate will tend to be more robust than one which was struggling in the first place. Some assays, such as two site or coated plate methods, will always be more precise because it is easier to achieve conditions of reagent excess.

In this chapter I will only describe the main categories of assay and discuss their relative merits. Novel assay procedures, often incorporating special equipment, are covered in enzyme-mediated immunoassay by Ngo and Lenhoff and ELISA by Kemeny and Challacombe (see bibliography). I have limited this chapter to microtitre plate, tube and pin assays. The assay designs I shall describe apply equally to all three systems and therefore I shall just refer to the commonest of these, microtitre plates. In the interests of space, specific methods are not described in detail and washing and other routine steps are not mentioned. These steps are described in detail in the appendix.

NON-COMPETITIVE

Coated plate

Antigen-coated plate

These are probably the simplest type of assay. They are also called indirect ELISA, dirty plate assay and sandwich ELISA. Microtitre plates are coated with antigen (directly or via some coupling agent - see chapter 4). Sample is added and the bound antibody subsequently detected by addition of an enzyme-labelled antibody specific for the bound antibody (Fig 3.1). This enzyme labelled antibody is referred to as the detector antibody. The detector antibody is not always labelled directly and a second enzyme-labelled anti-globulin antibody, directed against the detector antibody, is sometimes used.

For example, the detector antibody may be an expensive monoclonal antibody. Under these circumstances the answer may be to use an additional antibody directed against mouse immunoglobulin. Although it is generally best to use the simplest method, this extra step cuts down on the number of enzyme labelled reagents that we have to monitor. Others use the biotin/ avidin system in much the same way. The detector antibody can be biotinylated, biotin binds specifically to avidin with a very high affinity, so; enzyme-labelled avidin which binds to the biotinylated antibody is used. Possible short circuits and other difficulties are discussed in chapter 7.

Antibody-coated plate assays

Two-site

The two-site assay exemplifies some of the best features of ELISA. Some-times referred to as a sandwich assay, the two-site ELISA uses a pair of

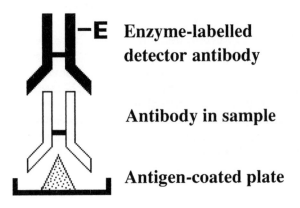

FIG 3.1. Antigen-coated plate ELISA.

(a)   **(b)**

Enzyme-labelled   **Enzyme-labelled detector**
detector antibody   **antibody to determinant 1**

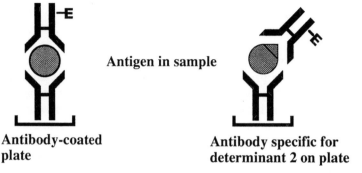

Antigen in sample

Antibody-coated   **Antibody specific for**
plate   **determinant 2 on plate**

FIG 3.2.The two site ELISA. (a) Symmetrical, (b) asymmetrical.

antibodies, one bound to the plate and the other labelled, to first bind and subsequently identify the presence of an antigen. They may be symmetrical (Fig 3.2 a) if the same reagent is used for capture and as detector, or are termed asymmetrical where these are different (Fig 3.2b). It is essential that a high proportion of the antigen in the sample binds.

Symmetrical assays

In this type of assay the microtitre plate is coated with antibody to the test antigen. The test antigen is added and allowed to bind to the antibody on the plate. Bound antigen is subsequently detected by addition of the same antibody that was used to coat the plate but which has been enzyme-labelled. It is not possible to use a second enzyme-labelled antibody as this would react with the coating antibody too although it is possible to biotinylate the second antibody and follow this with enzyme-labelled avidin. Symmetrical assays usually utilise polyclonal antisera because it is rare (although not unheard of) for the same monoclonal antibody to be suitable both for capture and as detector.

Asymmetrical assays

Asymmetrical assays have become increasingly popular as affinity-purified antibodies are expensive to prepare in commercial quantities. One of the best combinations uses a monoclonal capture reagent and a polyclonal detector.This has the advantage of providing a high degree of specificity during antigen capture and maximises sensitivity by allowing a large number of enzyme-labelled detector molecules to bind to each molecule of antigen, as the polyclonal reagent will recognise a number of different epitopes on the antigen. It also makes it possible for second, enzyme-labelled antibodies to be used. The procedure is essentially the same as for symmetrical assays except that the capture and detector antibodies are different. However, monoclonal antibodies are not always good capture reagents compared with polyclonal antibodies which usually perform well. There will, therefore, be circumstances where it is necessary to use an affinity-purified polyclonal capture antibody with a monoclonal detector.

Class capture

There are times when it is desirable to separate the sample from the medium in which it is found prior to carrying out the assay. This is not limited to immunoassays and pre-extraction of the sample is used in many chemical tests. In class-capture immunoassays the immunoglobulin class of interest is removed from the other immunoglobulin classes. These might contain antibodies that could interfere with the reaction between the sample antibody and antigen. The immunoglobulin class for which this is the greatest problem is IgM.

IgM

IgM antibodies are diagnostic of recent infection with a number of pathogens, such as rubella. The difficulty is that there may be many times more IgG anti-rubella antibodies, which are always present regardless of infection and so are of no diagnostic value. The problem is compounded by the fact that the IgG antibodies present will tend to be of much higher affinity and so may disproportionately inhibit binding of IgM antibodies to the antigen. A similar problem exists when one is trying to measure IgG subclass antibodies where antibody of one subclass may be of much higher affinity than that of another.

In IgM class-capture assays the microtitre plate is coated with polyclonal or monoclonal antibody specific for IgM (Fig 3.3). This anti-IgM-coated plate is used to absorb most of the IgM in the sample. Any IgM antibody activity bound to the plate is determined by addition of labelled antigen. The efficient capture of immunoglobulin of the appropriate class is essential in this assay and this can be checked in two ways: either with radiolabelled immunoglobulin or with labelled antisera specific for the corresponding immunoglobulin class.

IgE

IgE antibodies are usually found in serum or other body fluids in the presence of an excess of IgG antibodies which can interfere with their detection. IgE antibodies, however, are not usually of low affinity. Class-capture assays for IgE have been developed but in most cases it is easier to assay the samples at a dilution where there is sufficient antigen to bind antibodies of both immunoglobulin classes.

Class-capture can be used to determine the absolute quantity of IgE by saturation with labelled antigen. Here IgE is bound to anti-IgE-coated removable microtitre plate wells. Increasing amounts of ^{125}I radiolabelled

Enzyme-labelled antigen

Immunoglobulin class (eg: IgM or IgE)

Anti-Ig coated plate

FIG 3.3.Class capture ELISA.

antigen are added to successive wells until a plateau of binding is reached. At this point it is assumed that there is a ratio of antigen to IgE of 2:1. From the known molecular weight of the antigen, and its specific radioactivity, the absolute amount of antigen bound can be determined and from that the precise amount of IgE antibody.

COMPETITIVE

As discussed above, competitive assays make it possible to obtain an estimate of the amount of a particular antibody or antigen, even when these cannot be isolated from the medium in which they are found.

They can also provide useful information about the presence of common and distinct antigenic determinants. There is no doubt that these assays work best when the competitive part of the assay is carried out in solution, although this is not always possible. Here there can be an advantage in using coated pins which can be added to the microtitre plate containing the labelled antigen and sample, once the competitive phase is complete (Fig 3.4).

Antigen

Antigen can be measured in a competitive ELISA in two ways. In the first the microtitre plate is coated with the same antigen or antigen mixture and enzyme-labelled antibody specific for the test antigen added together with the sample. In a modification of this method the enzyme-labelled antibody and sample are incubated together before being added to the antigen coated plate, which is slightly more sensitive. If the corresponding antigen is present in the sample, the enzyme-labelled antibody will be prevented from binding (Fig 3.5a). An alternative system utilises an antibody-coated microtitre plate. Here, enzyme-labelled antigen competes with the unlabelled antigen in the sample for the antibody on the plate (Fig 3.5b). The drawback to the antigen-coated plate method is that a precise, limiting quantity of antigen must be bound to the microtitre plate. Furthermore, any dissociated antigen will also act as an

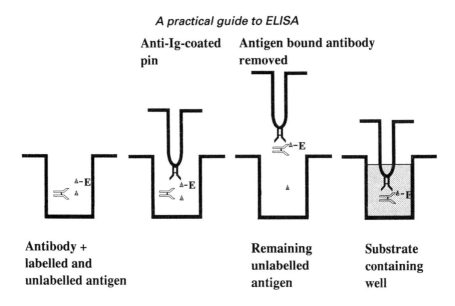

Anti-Ig-coated
pin

Antigen bound antibody
removed

Antibody +
labelled and
unlabelled antigen

Remaining
unlabelled
antigen

Substrate
containing
well

FIG 3.4. The microtitre pin ELISA.

inhibitor. The drawback with the antibody-coated plate method is that the behaviour of the antigen may be altered by conjugation to the enzyme. One solution to this is to label it with a small molecule like biotin. Its really a question of what reagents are available.

Antibody

Competitive assays for measuring antibody have become more common. They are used to compare antibody specificity in studies of antigen structure with monoclonal antibodies, for example. In this procedure the microtitre plate can be coated with antigen (Fig 3.6a). The test antibody competes with a fixed quantity of enzyme-labelled antibody for binding to the antigen on the plate. Alternatively, the plate is coated with antibody (Fig 3.6b) and the sample and enzyme-labelled antigen added. Here the competition is between the antibody on the microtitre plate and the antibody in the sample.

IN SITU

Histochemical

The use of enzyme-labelled antibodies originated in histocytochemistry. I do not, however, propose to cover these techniques here. Instead I have included methods in which cell secretions are measured and those that involve the identification of proteins bound to nitrocellulose.

ELISA plaque

This technique has its origins in in the 'Jerne plaque assay' which detects

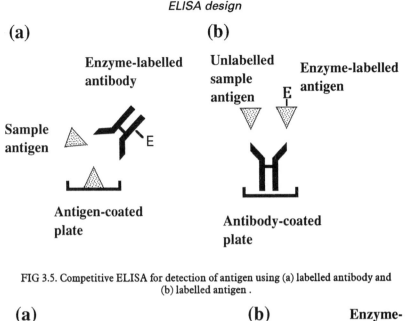

FIG 3.5. Competitive ELISA for detection of antigen using (a) labelled antibody and (b) labelled antigen .

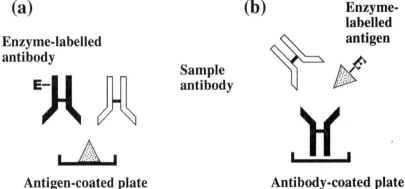

FIG 3.6. Competitive ELISA for detection of antibodies through competition with (a) labelled antibody and (b) labelled antigen.

individual antibody secreting cells by the lysis of antigen-coated red cells. Complement-fixing antibodies from antibody secreting cells lyse red cells nearby antigen-coated red cells. The clear spots formed are visible with the naked eye and are called plaques. The assays are described alternatively as the elispot and the ELISA plaque assays by the two laboratories in which they were developed.

Antibody secreting cells are cultured on plastic plates coated with antibody or antigen (Fig 3.7). The antibody secreted is bound close to the parent cell although it also extends some considerable distance from it. The antibody secreting cells are then washed away under non-physiological conditions so that they lyse. Enzyme-labelled antibodies to the secreted antibody which has bound to the plate are then added and finally agar containing the substrate 5-bromo-4-chloro-3-indoyl phosphate (BCIP) for alkaline phosphatase or p-nitrophenylenediamine for horse radish peroxidase. After a couple of hours blue spots can be seen (Fig 3.8), each spot corresponds to a single antibody-secreting cell. Optimum colour development is achieved after 24 hours. The spots can then be counted with the naked eye or under modest magnification.

We encountered considerable difficulties with the method until we used the same microtitre plates that we use for ELISA. It is not only antibody-secreting cells that can be detected in this way. Many secretory cells are amenable to this technique as long as sufficient protein is secreted. For example, lymphokine secretion by T cells has been measured in this way. The method is not quantitative but the size and intensity of the spot gives you some idea of the amount of a particular protein secreted by the cell. I suspect that there is probably an upper limit to the distance that a given protein can diffuse away from the cell and still bind intensely to the coated plate.

Dotblot

One of the simplest and yet the most practical applications of ELISA is the dotblot assay. In this technique antigen or antibody is spotted onto nitrocellulose paper using a fine pipette. The strip is then incubated with a solution of an irrelevant protein such as bovine serum albumin (BSA) or casein to block any vacant protein binding sites on the nitrocellulose before being incubated with enzyme-labelled antibody. The presence of bound enzyme is indicated by the development of colour with a substrate which produces an insoluble product, such as BCIP for alkaline phosphatase (Fig 3.9) and results in an intense blue spot. The method can be used to titrate antigen spotted onto the nitrocellulose or to detect antibody by addition of a second, enzyme-labelled, antibody.

The main virtue of this method is that nitrocellulose binds a wider range of proteins than plastic and in larger amounts. It is particularly useful for measuring monoclonal antibodies against antigens which are only available in crude form. When the concentration of the antigen in the extract is low it is possible to concentrate the sample onto the paper in a special washer (slot-blot).

Immunoblot

Also known as western blotting, in this technique proteins are separated by electrophoresis before being transferred to nitrocellulose paper. The strips are then incubated with enzyme-labelled antibodies and precipitating substrate. Positive results can be seen as a series of bands (Fig 3.10). It is essential to make sure that a sufficient quantity of the separated proteins is transferred from the acrylamide gel to the nitrocellulose paper. This can be done by staining the old acrylamide gel and by staining a sample of the nitrocellulose paper. The proteins can be separated in acrylamide or agarose gels. The former gives

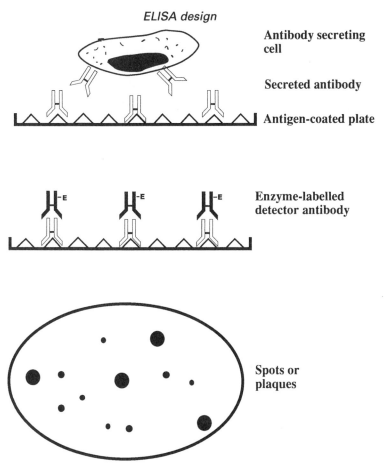

FIG 3.7.The ELISA spot or plaque assay.

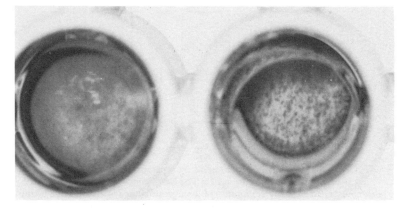

FIG 3.8. ELISA plaques of IgE secreting B cells.

better resolution; but with the latter it is possible to blot the proteins onto the nitrocellulose paper passively. Further resolution can be achieved by two-dimensional electrophoresis where proteins are first separated by isoelectric focusing (charge) and then by size (SDS polyacrylamide gel electrophoresis).

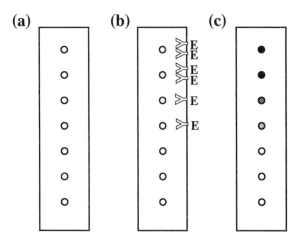

FIG 3.9. The dot blot assay. (a) Different amounts of sample are bound to the nitrocellulose strip. ((b) Enzyme-labelled antibody binds to the antigen spots. (c) An insoluble substrate is used to visualize the bound enzyme.

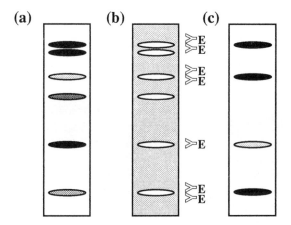

FIG 3.10. The western blot. (a) Proteins are separated by electrophoresis and transferred to nitrocellulose paper. (b) The bands containing antigen bind to enzyme-labelled antibody. (c) An insoluble substrate is used to visualize the bound enzyme.

CHAPTER 4

The Solid-phase Support and the Coating Antigen or Antibody

More problems in ELISA can be ascribed to the carrier used than to any other part of the assay. Yet without the adoption of microtitre plates it is doubtful whether it would have become such a widely practised technique. It is really a question of balancing convenience with the limitations of the particular solid-phase material. For example, in allergy where there may be as many as 200 different allergens it is not practical to use microtitre plates as there are inevitably some allergens that bind poorly to plastic. However, it is not possible to use the allergen-coated discs that are used for measuring IgE antibodies for IgG as the binding of IgG from unexposed individuals is too high. The answer is to consider what information you need and adopt the technique best suited to answering it. Indeed, you may not need a carrier at all and if, for example, you want to establish whether a new monoclonal antibody binds to a particular cell it may be simpler and more relevant to do immunofluoresence with the cell itself.

Carriers for ELISA can be divided into two broad categories, based on their capacity to bind protein. High-capacity materials include cyanogen bromide-activated agarose, cellulose and nitrocellulose. The chief advantage of these is that they can make use of relatively impure antigen preparations, and whole or partially pure antisera. In addition, they are stable, so that the coated carrier can be stored for months or even years. Their main drawback is that they are less easily washed and give higher background binding. Low-capacity carriers such as plastic and PVC are generally easier to wash and give low backgrounds but are more difficult to coat effectively.

The choice of carrier will depend on a number of factors. The most important of these is the purity of the coating material. This is especially true for measuring antibodies to crude allergen, viral or bacterial extracts. For example, if only one part in 1000 of an available extract of prawn is biologically relevant, it is still possible to use it to measure IgE antibodies if a high-capacity carrier is used. Where non-specific binding is a problem, in assays for IgG antibodies for example, a low-capacity carrier will be preferable. The range of the assay too is important. By virtue of their greater capacity, cyanogen bromide-activated carriers or nitrocellulose allow higher concentrations of sample to be measured without dilution.

HIGH-CAPACITY MATERIALS

Agarose particles and cellulose paper discs

One of the most significant advances in the development of sensitive solid-phase immunoradiometric assays (IRMA) was the discovery that agarose particles or cellulose paper could be activated by cyanogen bromide to bind a wide range of proteins. The reaction is inhibited by Tris salts - as I discovered by accident. Although not technically difficult, cyanogen bromide activation is hazardous and should not be undertaken by the inexperienced. Furthermore, if a small amount is needed (<15g) it is probably not worth preparing your own and is preferable to purchase it. Apart from being used in assays, antigen-coated agarose can be used to absorb cross-reacting antibodies and a wide range of animal serum and immunoglobulin coated-agarose beads are commercially available.

Although these high capacity carriers are stable for years, small amounts of protein are leached off and these can interfere with the assay. It is advisable, therefore, to wash them before use. When washing it should be noted that the briefest spin in the centrifuge is all that is needed. I have found that these particles, when coated with anti-immunoglobulin or protein A, are particularly useful immunoprecipitating reagents whose ability to precipitate is not affected by the concentration of the antigen or antibody (see equivalence in chapter 2).

Cyanogen bromide-activated cellulose paper discs are not available commercially but the discs are easily cut using an office paper punch and can be readily activated if the appropriate safety precautions are taken. These discs have the advantage that they are easier to handle and wash, but generally have a lower binding capacity than agarose particles.

Nitrocellulose paper

An alternative to cyanogen bromide-activated materials is offered by nitrocellulose paper. This is already activated and possesses a similar capacity to cellulose paper discs. Once protein has been bound, vacant binding sites are blocked with an excess of an irrelevant protein such as BSA or bovine casein. It is particularly useful in screening tests as a large number of samples can be spotted onto a small area. Washing steps are simple and quick to perform.

LOW-CAPACITY MATERIALS

Microtitre plates

A wide range of polystyrene microtitre plates (Fig 4.1) are available. Some are dedicated for use in ELISA while others are intended for other applications. The main advantage of using dedicated ELISA plates is that they vary little in their background absorbance and have a greater capacity to bind protein.

Some are available at different protein-binding capacities. They may come with information about IgG binding but for other proteins there is no alternative to testing different plates in your assay. When you do this it is important to use a range of concentrations of coating material, otherwise differences in binding may not be apparent.

Rigid plastic and flexible polyvinyl chloride (PVC)

These are the most widely used type of plate. They are easy to handle and are suitable for most ELISA assays. For most of our experiments we use rigid polystyrene plates. Before the advent of the newer polystyrene plates, flexible PVC plates were reported to have the greatest capacity to bind protein. Flexible plates have the additional advantage that they can be cut up with scissors and radioactivity counted in a gamma counter for measuring protein binding or in RIA. You can do this with rigid plates and a hot wire but it is very hard work!

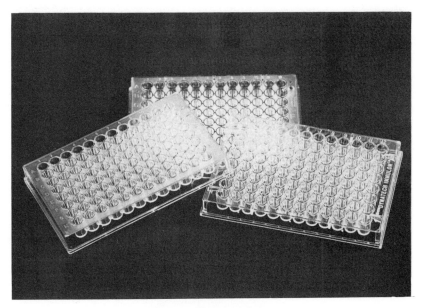

FIG 4.1. Microtitre plates

Removable wells and strips

This type of plate has the advantage that it is possible to mix differently coated wells in the same assay. They are particularly useful for kits where the user may only want to run a few samples. A wide range is available (Fig 4.2) and it is usually possible to purchase removable wells made of the same plastic as the rigid plates. They are more expensive and so should only be used where they offer a distinct advantage over rigid plates.

Pins

It is not always desirable to remove the sample for testing. For example, when screening cell culture supernatants for the presence of monoclonal antibodies it is possible to use sterile coated pins (Fig 4.3). The presence of antibody in culture supernatant can thus be detected without disturbing the cultured cells. Pin assays are claimed to be faster than those carried out in microtitre plates but I cannot see that lifting pins out of a microtitre plate is any quicker than flicking out a plate. I am aware of two types of pin that are currently available: the rigid 96 pin set up (Nunc, Denmark) and the snap fit FAST system (Flow labs, USA). Some can be sterilised by autoclaving while others may be soaked for a short time in 70% methanol. This treatment may, of course, alter the binding characteristics of the pins. In any event it is important to wash the pins with sterile culture medium rather than ELISA wash before putting them into the sterile cell cultures. In our hands the binding capacity of these pins is not as high as that of microtitre plates.

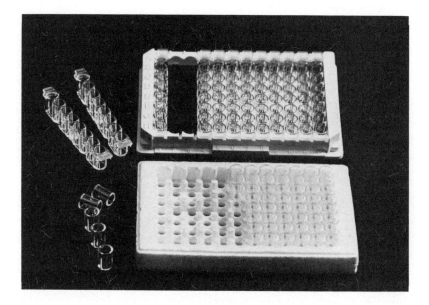

FIG 4.2. Removawells

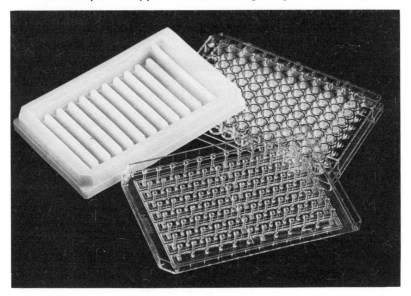

FIG 4.3. ELISA pins.

COATING CONDITIONS

After choosing the type of solid-phase to use, the next consideration is the procedure used for coating. For high-capacity carriers there are few pitfalls, so in the remainder of this chapter I will concentrate on microtitre plates. I have generally found that when there is effective coating, with a sufficient quantity of active material, the rest of the assay will fall into place.

Direct coating

By far the easiest method is to coat the plate directly by passive adsorption to the surface of the plastic. Optimal conditions will need to be determined, as described in chapter 5. Some proteins coat easily while others present difficulties. Direct coating is easier and should always be tried first. Two main factors govern the effectiveness of coated proteins: (1) the integrity of the bound molecules, which should not be damaged by binding to the plate and (2) the firmness with which they are bound. Both will be greatly influenced by the nature of the coating material. Proteins bound directly to plastic may be specifically orientated, for example through binding of hydrophobic regions to the plastic, so that some antigenic determinants fail to be expressed (Fig 4.4).

Desorption of the coated protein will adversely affect the assay (Fig 4.5). I once found that the capacity of an old preparation of human IgG to bind anti-human IgG antiserum was much greater than that of highly purified IgG myeloma proteins. I then discovered that the old IgG preparation were

ORIENTATION

FIG 4.4. Antigens binding to the plastic surface of the plate may become orientated in a particular way.

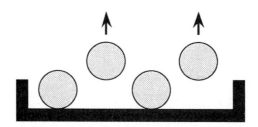

DESORPTION

FIG 4.5. Some of the antigen that is bound to plastic may become detached after a short time.

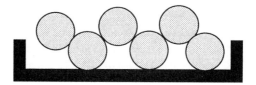

POLYMERISATION

FIG 4.6. Polymerisation of the antigen by aggregation, chemical modification or by antibody increases the number of points of attachment and so reduces desorption.

extensively denatured while the other preparation was not. Two things could have been different about the old IgG preparation which would certainly have contained aggregates, so that some of the IgG might have been held away from the surface of the plate and thus less likely to be altered by being bound to the plate. Furthermore, because it was aggregated, it is probable that there was multiple attachment to the plastic which increased the strength of attachment (Fig 4.6). The validity of this explanation has been established in a number of laboratories where antigens were polymerised with cross-linking agents such as glutaraldehyde and carbodiamide. In most cases, polymers performed better than native antigen.

36

Indirect coating

One solution to some of the difficulties of coating microtitre plates is to link the antigen or antibody to the plate via an intermediary. This has advantages in terms of reproducibility, sensitivity and specificity. The main drawback is that it requires more effort to set up and so should only be adopted if absolutely necessary. In addition, background binding is usually greater with these methods.

Indirect coating with antigen

Using antibody

One of the best methods of linking antigens to plastic is to bind them to plates coated with antibody (Fig 4.7). This has the added advantage of increasing the specificity of the assay, as contaminating antigens will not be able to bind.

The stability of the bound antigen is much greater for much the same reason as aggregated antigen - multiple attachment - (Table 4.1) and antigenic determinants are less likely to be altered through binding to the plastic.

If some of the linking antibodies are damaged they are unlikely to have any effect on the outcome of the assay - either they can or cannot bind antigen. There are a number of possible objections to this method of linking antigen to plastic. First, the link antibody might cover up some of the determinants on the antigen; secondly, antibody which becomes detached could interfere with the assay. In practice, the large number of antigenic determinants present on most proteins make either unlikely; indeed sensitivity is usually enhanced with these methods (Fig 4.8) suggesting that the advantages of better presentation and stable binding outweigh any possible inhibitory effects that might occur.

Both polyclonal and monoclonal antibodies can be used, although polyclonal antibodies are generally more predictable in terms of binding to plastic. These may need to be affinity purified although an immunoglobulin fraction (prepared by precipitation with ammonium sulphate or over a protein A or G column) can work. In any event, the capacity for antigen should be determined by radiolabelled antigen, enzyme activity or some other assay.

INDIRECT PRESENTATION

FIG 4.7. If antigen is bound to an antibody-coated plate it will be presented in a range of positions well away from the surface of the plastic.

37

Table 4.1.

Comparison of the dissociation of antigen bound to different types of microtitre plate. Not only do different types of microtitre plate bind different amounts of antigen (bee venom phospholipase A2; PLA2) it leaches off them to a differing extent. Plates coated with affinity purified rabbit anti-PLA2 bound more PLA2 and released less.

	Desorption after 3 hours		
Microtitre plate type	ng bound	ng desorbed	% desorption
A	5	2	40%
B	4	2	50%
B + anti-PLA2	31	0.4	1.3%
C	17	6	35%
D	2	1	50%
E	35	3	9%

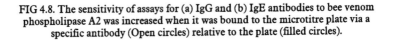

FIG 4.8. The sensitivity of assays for (a) IgG and (b) IgE antibodies to bee venom phospholipase A2 was increased when it was bound to the microtitre plate via a specific antibody (Open circles) relative to the plate (filled circles).

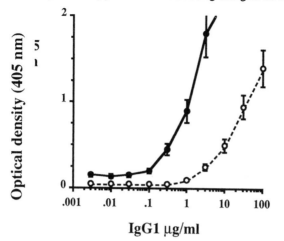

FIG 4.9. The capacity of mouse monoclonal anti-human IgG1 antibody for human IgG was greater if bound to an anti-mouse IgG coated plate than a plate (filled circles) coated directly with the monoclonal antibody (open circles).

Biotin/Avidin, Fluorescein, TNP

The method described above may not always be practical, as antibodies suitable for linking antigens may not be available. Also, there are circumstances when a number of different antigens need to be bound; then it may be preferable to use a common reagent to bind them all. Most proteins can be biotinylated and using avidin-coated plates, a wide range of biotinylated antigens can be bound. Similarly, plates coated with antibodies to fluorescein or haptens (low molecular weight chemicals than bind to antibodies) such as tri-nitrophenyl phosphate (TNP) are able to bind antigens labelled with these small chemical determinants. Of course the antigenic integrity of these modified reagents will need to be assessed.

Coating with antibodies

Most antibodies will retain their ability to bind antigen when coated to plastic, although there is little doubt that a substantial portion of this activity is lost. Furthermore, some monoclonal antibodies perform poorly once coated. Under such circumstances the antibody itself can be bound to the plate via a linking agent. Any of the methods described for antigen above will work but it is most usual to use anti-mouse antibody. An example of this is a monoclonal antibody to human IgG1 (NL16). When bound direct to the plate this performs poorly. Linking it to the plate with anti-mouse antibodies increases its capacity to bind human IgG1 considerably (Fig 4.9).

Another situation where such an indirect method of coating the plate might be considered is where the antibody is present at a low concentration in tissue culture medium. Here it might be impractical to purify the antibody. Coating with tissue culture supernatant containing 5% or 10% fetal calf serum would result in FCS rather than monoclonal antibody being bound to the plate. One solution to this problem is first to coat the plate with anti-mouse Ig.

PLATE MODIFICATION

There are a number of chemical processes in which the plate itself might be treated to increase its capacity to bind protein but these are beyond the scope of this book. It is, however, to be hoped that manufacturers will address the problem of binding adequate amounts of protein to their plates, both by looking at ways of incorporating reactive groups within the plastic and at least by furnishing the customer with the fullest details about the batches of plate supplied.

Glutaraldehyde

Possibly because glutaraldehyde was one of the first enzyme coupling reagents to be widely used, it was added to microtitre plates in an attempt to increase their capacity for protein binding. In our hands it is more common to find that antigenic activity is lost. An alternative is to treat the coating material with this first and then to use polymers to coat the plate. One might even use glutaraldehyde to link the coating material to plates coated with serum albumin or keyhole limpet haemocyanin (KLH).

COATING ANTIGEN OR ANTIBODY

If the most critical choice in ELISA is the type of carrier to use, the next most important must be the material used to coat the carrier and the method used to bind it. It is certainly important to know as much as possible about what you are coating and to store it in such a way as to retain its immunological integrity. In most cases coating materials are proteins. Proteins were once thought to be fragile molecules easily denatured by extremes of heat, pH and agitation. While this is undoubtedly true for some, many are in fact quite robust. The secret is to know your protein and handle it accordingly.

Antigen

Antigens require special consideration because they are so varied. It is essential to have a thorough understanding of their physicochemical nature and their biological role. Wherever possible it is preferable to use pure antigens but in many cases crude extracts are the only source of material available.

Cells

For antigens present on the surface of cells it may not be possible to isolate sufficient material to coat the plates. Under these circumstances the cells themselves can be used. Cells can be bound directly to plastic microtitre plates or indirectly with an appropriate antibody. A novel method of binding bacteria to plastic is to fix live cells in the plate with methyl glyoxal, a commonly used bacterial fixative. This *in vivo* method of binding the antigen also serves to increase the avidity of the cell for the plate as the cells are likely to form multiple attachments to the plastic which are then fixed.

Antigen mixtures

Very often the antigens have not been isolated and time may sometimes be better spent isolating them rather than developing assays. Using antibodies to select the appropriate antigens is one way of solving the problem. Alternatively the antigens can be separated by electrophoretic techniques. If all else fails one can bind the crude antigen extract to nitrocellulose, agarose beads or cellulose paper discs which can subsequently be incubated with the appropriate enzyme-labelled antibody.

Single antigens

Wherever possible it is best to use pure antigens. It may seem obvious, but it is always worth checking whether the antigen in question is available commercially. Even when specific antigens are not available commercially it may be possible to obtain them from other research workers. Obviously one should be realistic in one's expectations and some reagents are expensive to prepare or are only available in small quantities. A general search of the literature both within and outside one's own discipline can be rewarding. What may be an unknown protein to an allergist may be a common plant enzyme to the botanist. We were recently studying allergens in castor bean seeds when I discovered by accident that a group (in a different discipline at another institution) had already isolated a number of the most important proteins, which saved an enormous amount of work.

Peptides

It is possible to use peptide fragments of proteins and an ingenious system has been devised in which overlapping peptides (one to eight, two to nine, three to ten, etc) are synthesised onto pins. The binding of antibody to these peptides can be determined and the antigenic determinants mapped out. Of course it is only possible to define linear determinants by this method.

Antibody

Polyclonal

Absorption

Before the antibody is processed in any way it should first be absorbed against any other antigens with which it might bind. This usually means serum-coated agarose beads or other strains of bacteria or viruses. It is always more costly on antibody to absorb at later stages of purification when the antibody is less stable.

Whole serum

In many cases it is possible simply to coat the plate with a high-titre antiserum. If this works well then a great deal of time and effort can be saved. If not, the antiserum must be processed in some way.

Isolated immunoglobulin

It can be as much, if not more, work to prepare immunoglobulin as to affinity purify an antibody. However, some high affinity antibody may be lost during affinity purification and anyway the appropriate antigen may not be available in sufficient quantities for this. Under these circumstances it will be necessary to isolate the immunoglobulin fraction of the antiserum. Many methods have been described for this (Table 4.2).

It is not possible to go into the details of each of these methods here but they have been reviewed extensively elsewhere. I would suggest that the method should be tailored according to how precious the antibody is. In general, precipitation with ammonium sulphate followed by some form of electrophoretic separation works well. A number of immunoglobulin purification kits are now available.

Affinity purified

I have generally found that affinity purified antibodies work best as capture antibodies. They are the least stable form of antibody and as such should be handled with care, as described below. A range of eluting conditions can be used and these are detailed in the appendix. I generally start with the most gentle conditions such as pH 2.8 glycine/HCl and then move on to more vigorous conditions such as 0.001M HCl, and finally 6 M urea. It is essential that the activity of the antibody is monitored during affinity purification. This is most easily done by double gel diffusion.

F(ab')$_2$ fragments

There are some assays in which the antibody may react with the test material through other mechanisms. For example, the sample may contain rheumatoid factors (autoantibodies against immunoglobulin); to avoid interference by these it is necessary to use the F(ab')$_2$ fragment of the capture antibody (rheumatoid factors bind to the Fc portion of immunoglobulin). In addition, the ratio of signal to noise is usually better when F(ab')$_2$ rather than intact antibodies are used to coat the solid-phase. This may be useful in ultrasensitive assays.

Table 4.2.
Methods for isolating antibodies.

METHODS FOR ISOLATING ANTIBODIES	
Physicochemical methods	Affinity methods
Ammonium sulphate precipitation	Anti-immunoglobulin agarose
Ion exchange chromatography	Protein A
Chromatofocusing /Isoelectric focusing	Protein G
High performance liquid chromatography	

Monoclonal

Most of the problems of specificity that that are found with polyclonal antisera do not apply to their monoclonal counterparts. These do, however, present different problems simply because they represent a single example of the spectrum of antibodies that will be produced during an immune response. Antibodies with unusual properties are unlikely to be seen in a polyclonal mixture but once cloned they can be seen.

Nevertheless monoclonal antibodies can and do perform well. It is not, however, practical to absorb out cross-reactions as all the antibody may be lost. Of course, with both ascites and tissue culture supernatants there will be antibodies from the host animal or from the added fetal calf serum.

Ascites

Ascitic fluid collected following inoculation of a naive animal with a clone of antibody secreting cells can sometimes be used as it is. It is more usual, however, to prepare an ammonium sulphate fraction. Such immunoglobulin preparations can be quite pure but others may be contaminated with albumin which can subsequently be removed by addition of QAE Sephadex or agarose coated with anti-albumin.

Culture supernatant

As for ascites, whole culture supernatant can be used. The concentration of antibody is generally much lower, so it is less likely to work well as a capture reagent. It is also less practical to purify the mouse immunoglobulin. Either anti-mouse coated microtitre plates can be used or if the monoclonal is to be isolated then one of the affinity methods described above would be more suitable as they have the effect of concentrating the immunoglobulin. It is worth remembering that there will be much more bovine than mouse immunoglobulin. This can be removed, if necessary, by absorption with anti-bovine serum-agarose (negative affinity chromatography).

REAGENT HANDLING

Testing commercial products

It is much easier to use commercial reagents and it is therefore sensible to test available sources before preparing one's own. Some enlightened manufacturers recognise this and provide test aliquots free or for a nominal charge.

Storage

Temperature probably causes more worries than anything else and I think it is fair to say that there is no simple answer. We have a different policy for coating reagents than for detectors and generally store them frozen at -20°C. It is probably a good idea to freeze at -70°C first and then transfer to -20°C, as many proteins occur in eutectic mixtures (i.e. have different freezing and melting points), such as serum.

One way of reducing damage caused by freezing and thawing is to store in 50% glycerol. Recently we have found that this can freeze in our -20°C freezers and we have therefore increased the glycerol concentration to 75%. In some cases we do not add any glycerol but just store frozen. It must be remembered that there will be evaporation on storage and one should always wash out the aliquot rather than take out a fixed amount. If necessary make up too much and throw some away. We normally store coating antibodies and antigens in 10 µl aliquots with 20 µl of glycerol (these should be whirl mixed before being placed in the freezer). Freeze-drying is generally regarded as the most stable form in which to store proteins and for both antibodies and antigens we regard this as the best method for long term storage, particularly for 'sticky' proteins. I don't generally store coated plates but prefer to prepare them as and when they are needed.

CHAPTER 5

Enzyme-labelled Detector and Substrate

The low non-specific binding of enzyme-labelled reagents to plastic means that they can be used at a high concentration in ELISA, which places less demand on them in terms of affinity or purity, although the ultimate sensitivity of the assay will, of course, still be affected by the quality of these reagents. I don't doubt that in some cases better reagents can be made, but those available commercially are generally adequate for our purposes and cost less, in terms of time and effort, than producing them ourselves. Often I have heard people say how much cheaper it is to use a reagent at 1/1000 rather than 1/300. This is absurd - the difference in overall cost is trivial but the difference to the performance of the assay may be considerable. For more information I strongly recommend that you consult a recent book on ELISA by Tijssen (see reference section). In this chapter I will direct you to those methods and reagents that we have found to work well and that are simple to use.

ENZYME LABELS

The three enzymes most commonly used in ELISA are horse radish peroxidase (HRP), alkaline phosphatase (AP) and β-D-galactosidase (β-GAL) (probably in that order). While it may not be necessary to know everything about them it is important to have some idea about the conditions they require and what might interfere with them. It is important to measure enzyme activity under optimum conditions of pH and temperature. Immobilisation of the enzyme conjugate may alter the activity of the enzyme. It may increase enzyme activity; you may have noticed that colour sometimes starts to develop at the edge of the well. Reduction in fluid movement at the surface of the plastic may also lead to depletion of substrate locally.

DETECTORS

In immunoassay, a 'detector' is any molecule that has specificity for the sample and which is used to detect it. It can be labelled directly with enzyme or indirectly, as described below. In non-competitive assays the detector antibody or antigen should be used in excess. In practice this is not always possible so conditions as close as possible to this should be used. Where more than one detector is used, then each should be optimised in turn.

With high-capacity solid-phase carriers, like the allergen cellulose paper discs used to detect IgE antibodies, it is difficult to achieve saturating conditions for the bound IgE antibody. If the amount of labelled anti-IgE is increased, then the background becomes unacceptable. Because plastic is such an unsuitable carrier for many allergen extracts, which are only available in impure form, the solution is to use a high capacity carrier (cellulose paper discs) and a limiting concentration of labelled anti-IgE. At relatively low levels of IgE antibody the assay is more precise because the detector is nearer to excess and each sample can be diluted to fall in this more precise region.

The sample can be titred out which helps make sure that the point of measurement is close to the optimal assay condition, as in the dot blot assay, Titration of the sample involves additional work but is the only way of reconciling saturating conditions and high capacity carriers. Quantification and precision will be discussed in detail in chapter 7.

The specificity of the detecting antibody, however, is critical. Cross reactions between the detection antibody and other immunoglobulins can be neutralised by pre-incubation with a small amount of serum (1–2%) from the offending species or by pre-absorption with serum from that species bound to agarose. Cross-reacting antibodies cannot always, however, be absorbed out in this way and such antisera may be unsuitable for use in immunoassays.

Antibody

Direct vs indirect labelling

Detector antibodies can be labelled directly. This certainly makes for simpler assays but if the antibody is in short supply, or if you are new to the technique, it is probably better to use a commercial second antibody. If, for example, enzyme-labelled anti-mouse antibody is used then direct labelling of the monoclonal antibody will not be necessary. It also means that a single enzyme conjugated reagent can be used for many different assays. The concentration of this can, however, be critical and we have found that it is important for this reagent to be considerably in excess. Biotinylated antibody and enzyme-labelled avidin offer a similar advantage. For example, when measuring IgG antibodies we use a monoclonal anti-IgG (clone 8a4) and commercial rabbit anti-mouse IgG conjugated to alkaline phosphatase (Fig 5.1). This increases the number of steps and the risk of a short circuit (see chapter 7) but has been carefully optimised.

Fab' and F(ab')2

For extra sensitivity it is important to keep the background to a minimum. Removal of the Fc portion of the antibody (Fig 5.2) yields conjugates with

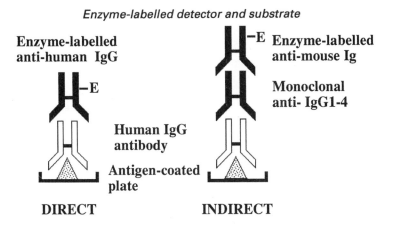

FIG 5.1. Enzyme labelled detector (DIRECT) compared with an enzyme-labelled antibody to the detector (INDIRECT).

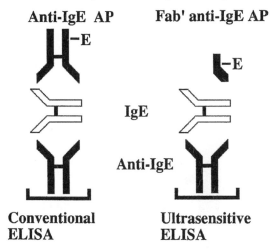

FIG 5.2. The enzyme labelled detector antibodies used in the conventional and ultrasensitive ELISA.

increased tissue penetration (for histology) and lower background binding. This is especially important when amplified substrate systems are used (see below).

Bridging antibodies

It is also possible to make use of the two combining sites on the detector antibody to form a bridge between the detector antibody and the enzyme-anti-enzyme complex (Fig 5.3).

Great sensitivity has been claimed for this method which also makes it possible to use a single enzyme conjugate for a number of assays. The main limitation of this method is the extra care needed in setting up the assay and the increased possiblilty of short circuits (see chapter 7).

RABBIT ANTI-
ENZYME/ENZYME
COMPLEX

GOAT ANTI-RABBIT
BRIDGING ANTIBODY

RABBIT ANTI-HUMAN
IgG DETECTOR

HUMAN IgG ANTIBODY

ANTIGEN-COATED PLATE

FIG 5.3. A bridging antibody can be used to link the unlabelled detector to an enzyme-anti-enzyme complex.

CHIMAERIC ANTIBODY

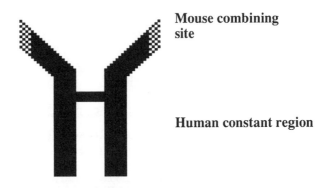

Mouse combining
site

Human constant region

FIG 5.4. A humanised monoclonal antibody which has a mouse combining site on a human immunoglobulin.

Chimaeric antibody

A chimaeric molecule is a hybrid. Chimaeric antibodies may have combining sites with different specificities or from a different species. These can be made by a molecular biological methods or by chemical means.

Monoclonal

Monoclonal chimaeric antibodies have been made which have mouse or rat combining sites incorporated into human immunoglobulin and vice versa. (Fig 5.4). These chimaeras have not proved easy to produce and so are only available for a small number of antigens.

Chemically prepared chimaeras

Chimaeric antibodies can be produced chemically, either by linking two antibodies (anti-enzyme and detector antibody) together with a cross-linking agent such as glutaraldehyde or by splitting the antibody into two halves with reducing agents and splicing one half onto another (Fig 5.5).

Antigen

Just as the discovery of monoclonal antibodies revolutionised the use of antibodies, so cloning antigens heralds an era in which events occurring at a cellular and even a subcellular level can be studied. The implications for those designing immunoassays are considerable and I strongly suggest that as these powerful tools become more widely available opportunities to improve existing assays or to set up methods that were previously impossible should be considered.

Direct versus indirect labelling

I know that there are circumstances where it is necessary to use enzyme-labelled antigens but I am always a bit apprehensive about it. The reason for my concern is that it is possible to assess the effect of conjugation on antibody function but it is much harder to do so for antigen. Wherever possible I favour an indirect approach in which the bound antigen is detected with an enzyme-labelled antibody (Fig 5.6). Alternatively, if the antigen must be labelled, small chemical groups such as haptens or biotin may have less effect on its ability to bind to antigen.

CHOICE OF ENZYME AND SUBSTRATE

The choice of enzyme and substrate will depend on a number of factors. In my view the most important of these is the consistency of results obtained. Substrates of horse radish peroxidase must be carefully prepared if consistent results are to be obtained. The rate of colour generation is also important. Horse radish peroxidase has a much faster turnover time (speed with which it

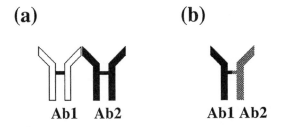

(a)   **(b)**

Ab1   Ab2   Ab1 Ab2

FIG 5.5. Chimaeric detectors can be formed chemically by linking: (a) two different antibodies together and (b) two different Fab' fragments.

(a)   **(b)**

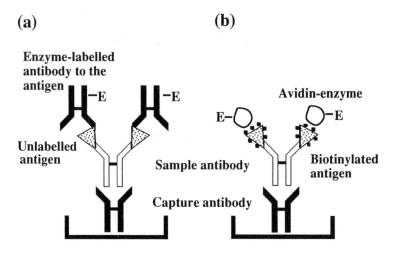

FIG 5.6. Antigen detectors can be prepared in three ways. The antigen can be directly labelled with the enzyme, as shown in figure 3.3; or (a) an enzyme-labelled anti-antigen antibody can be added after the antigen has bound; or (b) the antigen can be biotinylated and the enzyme coupled to avidin.

digests substrate) than alkaline phosphatase or β-D-galactosidase. Amplified substrate systems which have recently been described for alkaline phosphatase can overcome this as can fluorescent substrates for *β-D*-galactosidase. Finally, cost will be an important consideration. Horse radish peroxidase is easily the cheapest of the three most commonly used enzymes although some of its substrates are hazardous.

Alkaline phosphatase (EC 3.1.3.1)

Two sources of alkaline phosphatase are commonly used - *Escherichia coli* and bovine intestinal wall. Enzymes prepared from these two sources have different properties, with the bovine enzyme having a pH optimum of 10.3 while that obtained from E. coli has a pH optimum of 8.0. This is my preferred enzyme for ELISA. Conjugated to protein, it is stable for a long time and is resistant to bacteriostatic agents. The substrate, *p*-nitrophenyl phosphate (p-NPP), is easy to use and produces linear colour development with time. This substrate is stable, safe, and available commercially in convenient tablet form. Alkaline phosphatase is normally present at a low level in serum but is elevated in patients with liver disease and this can give rise to false positive results.

p-Nitrophenyl phosphate

Enzyme activity doubles between 25 °C and 37°C but hydrolysis of the substrate occurs at temperatures above 30 °C. The buffer which is usually used, phosphate buffered saline, contains substantial amounts of free phosphate ions (Pi). This is an inhibitor of alkaline phosphatase activity. Even shaking out the plates does not remove all the Pi as it stays in the unstirred layer - shaking out removes water not free phosphate. We now carry out all our assays in Tris/HCl buffer.

5-Bromo-4-chloro-3-indoyl phosphate

A number of substrates which yield insoluble products are available for alkaline phosphatase. One such substrate is 5-bromo-4-chloro-3-indoyl phosphate (BCIP). We have used this in the ELISA plaque assay and in the dot blot assay where it works well.

Amplified substrate system

The rate of colour development with alkaline phosphatase can be increased by using an amplified substrate system (Fig 5.7). In this the product of the first substrate catalyses the action of a pair of enzymes on a second substrate. The arrangement is rather like a high-fidelity amplifier, in which the small amount of signal from the cartridge is amplified by a small but precise amount by the pre-amplifier, and this larger, equally accurate, signal is then amplified by the power amplifier to a level where it can drive the loudspeakers. According to this analogy the cartridge is the alkaline phosphatase bound in the assay, the pre-amp is the NADP generated and the power amp is the redox cycle between alcohol dehydrogenase and diaphorase.

Of course if the background is high that will simply be enhanced. It is therefore necessary to use a detector antibody / enzyme conjugate which produces low background. This can be accomplished with the Fab' portion of the antibody linked to alkaline phosphatase at a 1:1 ratio (Fig 5.8).

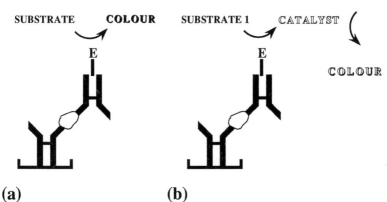

FIG 5.7. The amplified substrate system for alkaline phosphatase

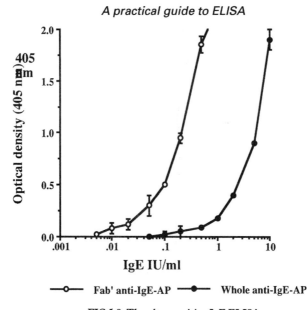

FIG 5.8. The ultrasensitive IgE ELISA.

Horseradish peroxidase (EC 1.11.1.7)

o-Phenylene diamine

This is probably the most widely used chromogen for horse radish peroxidase. The rate of generation of colour is not always linear, usually if an incorrect amount of hydrogen peroxide substrate, has been used. The pH too is critical and this can change when the chromogen salt (OPD) is dissolved in the substrate buffer. The hydrogen peroxide concentration is critical. Too much and the enzyme is inactivated, too little and sensitivity is lost. Optimal conditions are given in the appendix but each laboratory should test individual batches to confirm optimum working conditions in terms of sensitivity and linearity of colour development. Hydrogen peroxide can be stored in aliquots at 4°C; alternatively, urea peroxide tablets can be used.

Other chromogens

Another commonly used chromogen is 2,2'-azino-di-(3-ethyl-benzthiazoline sulphonate-6) (ABTS) which gives a dark, visually readable end point. Other substrates that have been used include O-toluidine, 5-amino salicylic acid (5-AS) and O-diansidine. Of all the chromogens, however, OPD is reported to be most sensitive at 492 nm but when compared at their optimum wavelengths (OPD 475 nm, ABTS 414 nm, 5-AS 500 nm), they are equally sensitive. 5-AS is poorly soluble, but solubility can be increased by recrystallisation in the presence of hydrogen sulphite. Two promising new substrates, 3- (dimethylamino) benzoic acid (DMAB) and 3-methyl-2-benzothiazoline hydrazone (MBTH) may help overcome some of the disadvantages of other HRP substrate chromogens. Fluorogenic substrates have been reported for peroxidase but these are less stable compared with those for other enzymes such as alkaline phosphatase and β-galactosidase.

β-D-Galactosidase (EC 3.2.1.23)

The cleavage of o-nitrophenyl-β-D-galactoside (o-NPG) to o-nitrophenol (o-NP) can be used to measure the activity of β-D-galactosidase. The concentration of o-NP is measured at 405 nm. Although it has a somewhat slower turnover rate compared with horseradish peroxidase and alkaline phosphatase, β-D-galactosidase has the advantage that it is not normally found in plasma or other body fluids, although it can be found in some micro-organisms. Coloured substrates such as p-nitrophenyl-β-D-galactoside, as well as stable fluorogenic substrates like 4-methylumbelliferyl-β-D galactoside (MUG), are available. With MUG, as little as one attomole (10^{-18}) of enzyme can be detected per hour. β-D-galactosidase can be coupled to proteins using the one-step glutaraldehyde and maleimide procedures.

Other enzymes and substrates

A number of other enzymes such as urease, glucose oxidase and ribonuclease, have been used, but less commonly, and so exceed the scope of this book.

METHOD OF CONJUGATION

A large number of different procedures have been described for linking enzymes to proteins. Ideally all the enzyme and detector protein should bind, the conjugate should contain defined amounts of each, there should be little or no inactivation of enzyme or antibody/antigen and the product should be stable during storage. Three methods of conjugation exist: chemical, bridging antibody and through other high affinity reactions such as biotin/avidin.

Chemical coupling

This is the most common method of coupling enzyme and antibody/antigen used in ELISA. It is worth bearing in mind that most of the reactions obey the law of mass action, which means that the efficiency of coupling is much greater the higher the concentration of the reactants. For example, lowering the concentration of antibody in the glutaraldehyde method from 5 to 1 mg/ml reduces coupling efficiency 25 fold!

Glutaraldehyde

The most widely applicable and simplest procedure for coupling enzymes to proteins is the one-step glutaraldehyde method. It is highly efficient (60 to 70%) and is simple to perform. However the conjugates formed tend to be of high molecular weight and are heterogeneous, with different numbers of enzyme molecules coupled to the antigen or antibody.

Some enzymes, such as horseradish peroxidase, cannot be conjugated very well using this method. For this reason a two-step glutaraldehyde procedure was developed in which the protein is reacted first with glutaraldehyde and then, after dialysis, with the enzyme. The coupling efficiency is lower (5 - 10%) but the majority of the conjugates formed have a 1:1 ratio of the enzyme to protein. These conjugates have the advantage that they give lower background binding which is essential for very sensitive assays.

Periodate

This is probably the preferred method for horseradish peroxidase. The carbohydrates (about 20%) of horseradish peroxidase are oxidised with sodium periodate, producing aldehyde and carboxyl groups. The aldehyde groups then form Schiff bases with the amino groups of the added antibody or antigen.

Maleimide

This method produces very high quality alkaline phosphatase and β-D-galactosidase conjugates. It uses a heterobifunctional reagent which forms a link between enzyme and antibody/antigen in much the same way as glutaraldehyde, but the well controlled sequential reactions used avoid undesirable cross-linking.

PURIFICATION OF CONJUGATE

For most labelled conjugates, the product will perform better if unbound enzyme and detector are removed. Indeed, this is essential for the periodate and two-step glutaraldehyde methods where a minority of enzyme molecules have been bound. A wide range of different chromatography media are suitable. The choice will depend largely on the facilities at your disposal. You will certainly obtain better results from a method at which you are skilled than with one with which you have little experience. When dealing with small amounts of conjugates such techniques as high performance liquid chromatography (HPLC) and fast protein liquid chromatography (FPLC) are ideal. If you don't have such facilities, gel filtration and ion exchange chromatography can be used. It is sensible to keep the size of the column to a minimum to reduce loss of sample. Gel filtration columns can be equilibrated with protein-containing buffers to reduce losses. In my hands, acrylamide based resins such as the AcA series from LKB and the Biogel-P series from Bio-Rad work best.

STORAGE

The yields of affinity-purified conjugates may be low. In such circumstances it is important to add serum from the same animal species or some other protein to the conjugate storage buffer to prevent loss of enzyme-labelled antibody during storage. Apart from antigen-affinity chromatography, horse-radish peroxidase conjugates can be isolated by the lectin concanavalin A (con A) coupled to agarose and eluted with 0.1M methyl-D-glucopyranoside.

Peroxidase is sensitive to microorganisms and to anti-microbial agents such as sodium azide and methanol and can be inactivated by plastic, although the latter can be prevented by the addition of Tween 20 to the diluent. Peroxidase conjugates should, therefore, be stored at -20°C in 75% glycerol and all buffers and wash solutions should be free of azide.

Alkaline phosphatase conjugates are normally stored at 4°C in buffer containing 1-2% w/v protein, 75% glycerol and azide and can, under these conditions, be stored at -20°C.

CHAPTER 6

Quantitation

WHY QUANTIFY?

If you want to compare the amount of a particular substance in different populations, be they people, animals or cells in culture, you will need to collect quantitative data. For example, most people make IgG antibodies to food proteins. The difference between people with a gastrointestinal condition such as coeliac disease and the rest of the population lies in the amount of antibody that they produce rather than its presence or absence. To obtain this information it is necessary to quantify the assay.

Quantitation is one of those subjects that tends to make most peoples' eyes glaze over (mine included sometimes). Yet it is a considerable source of error. We all accept that we live in a physical world and depend upon such units of measurement as the metre rule (UK & USA yard) and the kilogram (UK & USA pound). As scientists in the field of biology we should be no less concerned with the units of measurement specific to our own chosen field. That doesn't mean we need to be obsessive, but there are simple rules that should be followed. I have attempted to describe these in this chapter.

The ways in which we should express the results of ELISA are fairly straightforward, yet time after time they are ignored. Part of the problem is that there are no standards for many of the things that we measure. Many researchers express their data as optical density units in much the same way as they used to express it as counts per minute or % binding when they did radioimmunoassay and I suppose that they can at least comfort themselves with the knowledge that as measurements optical density and radioactive counts are real and that twice as many counts were just that. The trouble is that the relationship between the concentration of what they were measuring and the amount of colour generated is not a simple one and may in fact vary from day to day. However, without wishing to offend aficionados of immunoassay, quantitative aspects are shrouded by an alien jargon and incomprehensible mathematical formulae. The aim of this chapter, then, is to offer a simple approach to expressing data collected by ELISA.

FIXED CUT-OFF

For many applications it is sufficient simply to record those wells in which the colour developed beyond a preset value. These assays should be developed just as carefully as those which yield quantitative results. It is just that the user only needs a "yes" or "no" answer. There is an inherent danger that the lack of quantitative measurement will mean that reference materials are not run and that the assay itself may not have been optimised properly. Rather like the short story, these yes/no type assays need to be just as carefully developed as a full scale novel.

REFERENCE CURVES

There is one golden rule: 'You can't collect quantitative data if you didn't run a reference curve at the same time as the samples'. Remember - if you do have a reference curve you can choose not to use it but not vice versa.

What are they?

To many, constructing a reference curve is a rather daunting prospect. Let us first assume, for a moment, that there is a simple, linear relationship between the amount of colour generated and the concentration of the sample being measured. If the results were plotted they would look something like

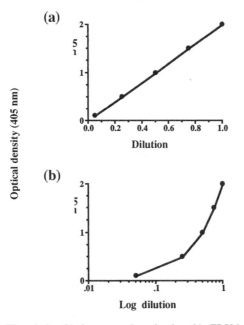

FIG 6.1. The relationship between colour developed in ELISA and sample concentration. If there were a 1:1 relationship between concentration and colour the appearance of the standard curve would be a straight line (a) if plotted on linear/linear graph paper or a curve (b) if plotted on semi-log paper.

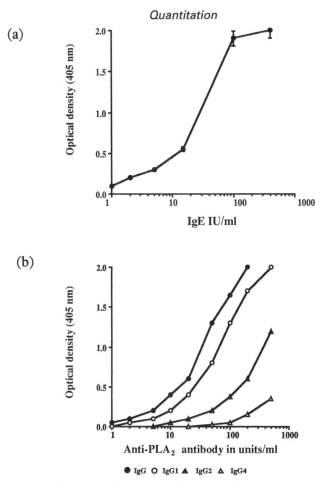

FIG 6.2. Typical ELISA curves for (a) IgE and (b) IgG subclass antibodies.

figure 6.1a. Please notice that the same data gives a curve (Fig 6.1b) when plotted on semi-log paper (as is usual).

For some of our assays this is indeed what we get (Fig 6.2a & b) but this is not always so. There are many reasons why a reference curve may not show a linear relationship between colour and concentration. First, the assay itself may have some deficiencies, e.g. too little antigen on the plate, too little detector. Secondly, the sample itself may be heterogeneous. For example, in a polyclonal antibody response there may be a mixture of antibodies of widely differing affinity. This is discussed in greater detail under 'parallelism' below.

It is because the relationship between the amount of colour generated and the concentration of the sample is not usually linear that a reference is needed. Because there will be a range of different concentrations of sample, a single reference point is insufficient. The choice of reference material and standards are discussed below.

How to plot

It is usual to plot the results on semi-logarithmic paper because most reference curves cover a wide concentration range. The reason for not choosing log log paper is that the intensity of colour generated is linear and covers a narrow concentration range (effectively limited to an optical density of 2.0-2.5) by the plate readers. With semi-log paper it is possible to plot the reference curve as units or as a weight concentration (i.e. mg, μg, ng per ml) using the scale provided. I am often asked how to plot a series of dilutions on this type of graph paper. The answer is as a series of reciprocal values (Fig 6.3) such that a 1/10 dilution is plotted at 0.1, 1/20 at 0.05, and 1/50 at 0.02 etc.

Results are read by taking the mean optical density (OD) of the sample and looking across to the reference curve. At the point on the reference curve corresponding to this value one should look down to the scale on the horizontal axis and record the value. This will then need to be multiplied by the appropriate dilution factor. It may seem obvious but the test sample must be run at the same volume as the reference sample. This method of recording the results makes an important assumption, that the sample and the reference give parallel dilution curves. This is discussed further below.

What shape curve should you get?

I can clearly recall being shown reference curves extending over a 4 or 5 log range of dilutions. The experimenter was very proud of the fact that the duplicates for the standard were so close and that a wide range of sample concentrations could be tested in a single assay. In fact it is not desirable to have a very shallow dilution curve because it means that a large change in

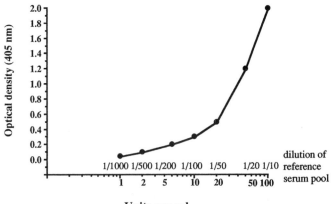

Units per ml

FIG 6.3. Calibration of the reference curve. For convenience the results are plotted on semi-log graph paper. Assuming the reference has 1000 units/ml then the value 1 corresponds to a 1/1000 dilution and a value of 20 to a 1/50 dilution. The results can be calculated by comparing the amount of colour in the sample wells against the curve and reading the units on the horizontal axis.

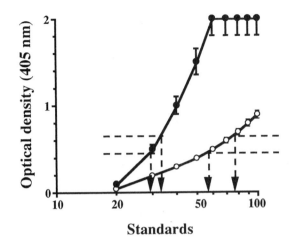

Standards

FIG 6.4. The effect of the slope of the reference curve on the precision of the assay. In the example shown above the same replicates give much less variation on the scale (horizontal axis) with the steeper curve (closed symbols) than the shallower one (open symbols).

concentration will only give a small change in colour (Fig 6.4). Such a curve usually means that something other than the sample is limiting the assay.

Where the slope is shallow even apparently good duplicates give a widely differing value while poorer duplicates give much less variation when read from a steep standard curve. This sounds a bit of a cheat but please bear in mind that the steepness of the standard curve reflects how well the assay is working. In practice dilution curves in ELISA should not extend beyond a 2 to 3 log range, with a reader giving a cut off at an optical density of 2.0.

What is the lower limit?

The lower limit of the assay always presents problems in ELISA. In our laboratory we use a lower limit of 1.5 times the background value. The background may be defined in two ways: either from a well containing all reagents except the sample; or, and preferably, from a well containing a known negative sample. It is important not to zero the plate reader on these wells as a poor assay may then go unnoticed. If the background is high then the working range of the assay will be inadequate and the assay will clearly require further development.

The effect of affinity

Affinity, as defined in chapter 1, is the strength with which antibody binds antigen. Both the affinity of the antibody (Fig 6.5) and the valency of the antigen can influence the slope of the dilution curve and it is difficult to know if one is dealing with a large amount of low affinity antibody or a small amount of high affinity antibody. Of course the affinity of the antibody will not be the

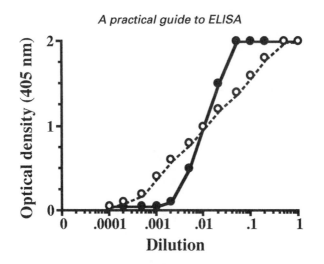

FIG 6.5. The effect of antibody affinity on the slope of the reference curve. A small quantity of high affinity antibody gives the type of curve shown by the solid line while a large amount of low affinity antibody will give a shallower curve as shown by the broken line.

only factor that determines the slope of the curve. There are a number of different ways of determining this and these are cited in the bibliography.

TITRE

This was the first form in which immunoassay data were collected. It is the lowest dilution of the sample that can be detected at a predetermined cut-off. Originally used to score red-cell agglutination assays, titres can provide a considerable amount of information about the sample. Unlike using a reference curve, where values will be recorded over a range of antibody/antigen ratios, the end point titre will have the same antibody/antigen ratio. Inevitably this means that the microtitre plate-bound component of the assay will be as close to excess as it is going to be. Thus, differences in affinity will be minimised and the result closer to the actual concentration of the sample.

For example, we recently found that human IgG1 antibodies to a bee venom protein, produced by children allergic to bee stings who had been given two years of immunotherapy, had an affinity nearly 100 times lower than corresponding IgG4 antibodies. When the samples were assessed in our optimised ELISA it appeared that there was a similar difference in the concentration of these two types of antibody. Compared using end point titre, however, the difference was only 10-fold, suggesting that part of what we were measuring was the effect of antibody affinity.

Because of the small amounts of effective antigen used in antigen-coated plate ELISAs I am convinced that affinity is a major influence. Indeed, when one tries to assess how much antigen is needed we have found that if presented on an antibody as little as 1 ng per well yields maximum IgG binding while 1000 times as much is required in direct coating. Of course this large difference is due partly to the particular antigen used but it shows how inefficiently antigen may be presented on these plastic plates. This is well illustrated by reincubating sample in a fresh well. Much of the original antibody activity remains. Fortunately this is not such a problem with antibody-coated plates.

Choice of end point

How is the end point chosen? For the red cell agglutination assays this was the lowest concentration at which there was no red cell button. Undoubtedly there was still antibody present as more sensitive versions of the assay have subsequently shown. Thus the end point is not the least amount of antibody or antigen present, it is the smallest quantity that can reliably be detected by the method used. End points that are too close to the background are unreliable because very small differences in concentration give equally small changes in colour. We normally use an end point 0.2 OD units above background (wells with diluent alone).

REFERENCE MATERIALS

Preparation of reference material

Apart from cases where there are internationally agreed standards it will be necessary to prepare your own 'in house' standard. This should ideally be a pool of high-titre sera or a high concentration of antigen. If appropriate, it should be freeze-dried although for many situations freezing at -20 °C or -70°C may be more practical. The reference material should be stored in suitable aliquots. We store our reference serum pools in two amounts: a small aliquot, which is sufficient for a single reference curve, and in larger volumes (usually 1 ml). It is a good idea to make sure that all the material has been frozen and thawed the same number of times. The question of adding glycerol and other aspects are the same as for handling coating reagents described in chapter 4. It is important that both positive and negative reference material is prepared in this way. An independent sample should also be included as a quality control (see below).

Assigning units

To assign units we look at the dilution that gives an optical density of 0.2 and determine the titre of the sample. The units assigned are proportional to the nearest log10 titre. Thus a titre of 1/10,000 would be assigned 10,000 arbitrary units. In this way it is possible to gain some idea of the relative amount of antibody present in different assays but such comparisons are, of course, only semi-quantitative.

Standards

Why have standards? Primarily so that results obtained in different laboratories within and between different countries can be compared. Why don't we just swap reagents between laboratories? Certainly this is better than nothing and for many substances it is the only practical approach. Eventually, as more and more people want to use a particular reference preparation, it becomes impractical to do this properly without some sort of organisation. Furthermore, if all the laboratories who are swapping reference reagents are not expert at accurately dispensing and handling standards then the data may be uninterpretable. Wherever possible, it is desirable that internationally agreed standards should be used.

WHO/IUIS

The World Health Organisation (WHO) and the International Union of Immunological Societies (IUIS) are the two bodies responsible for immunological standards. Because the reagents will be used worldwide it is important that they get it right. It is also important that sufficient primary standard is prepared to satisfy demand for 20 years or more. The development of new methods will invariably call for the periodic revision of standards. For this reason it is not a good idea to give absolute values for standards but rather to assign units, the value of which can be can be revised later if necessary. For example, the international standard for IgE is approximately equal to 2.4 ng. Remember that the primary function of internationally agreed standards is that they enable someone working in one laboratory to compare their values with those obtained in another.

A more difficult question is whether the units for related compounds should bear any relation to one another. The answer is, I think, a compromise. An approximate relationship to the quantity of the particular material is desirable. The end user will do this anyway so it is probably better that it is done properly.

Commercial standards

The widespread use of commercial kits means that their calibrants play an increasingly important part in standardisation. For example, in the absence of standards for human IgE antibodies the reference material sold by Pharmacia has made it possible to make approximate comparisons between different centres. However, some commercial companies may try to promote their own reference material over WHO/IUIS standards which seems to defeat the object of the exercise.

Working standards

It is only possible to make a limited quantity of a standard. Primary standards should, therefore, only be used to calibrate a working standard prepared in house. This should be handled in much the same way as the reference reagents described above. In some cases it is possible to buy calibrated working standards.

MONITORING PERFORMANCE

Quality controls

When I first carried out immunoassays I used to try to run all the samples in the same assay. This is obviously not practical if there are a large number of samples. In one sense I was right to do so because some of those early assays were quite variable. In order to check that comparable results are obtained in different assays it is necessary to run quality controls. These are simply samples which are run in each assay. They should be run at at least three dilutions to cover the upper, middle and bottom parts of the standard curve (Fig 6.6). If the result obtained with these falls outside acceptable limits then the assay will need to be repeated. Acceptable limits will vary for different assays but is typically within 10, 15 or 20%.

Coefficient of variation

By running quality controls in each assay it is possible to determine the coefficient of variation (CV). This is a measure of the variation between assays and the formula for this is:

$$CV = \frac{\text{Standard deviation}}{\text{Mean}} \times 100$$

Intra (within)-assay and inter (between)-assay variation should be monitored. At least four or five assays need to be run before the coefficient of variation can be properly determined. With new batches of reagent one

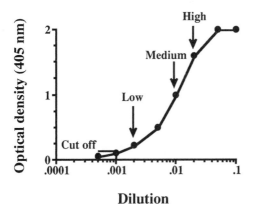

FIG 6.6. A reference curve showing the points at which standards would normally be run and the lower cut off.

sometimes sees a jump in the CV so if comparisons of the data are to be made it is a good idea to try and use the same batch of reagents throughout. One extra advantage of monitoring variation in this way is that it allows one to be confident about accepting or rejecting results produced by new members of staff. The limits you decide to accept or reject are up to you and may differ from assay to assay. The simplest way of expressing quality control values is simply to plot them week by week (Fig 6.7).

Parallelism

Throughout this chapter the assumption has been made that it is acceptable to record results from a reference curve. This is only true if the sample dilutes

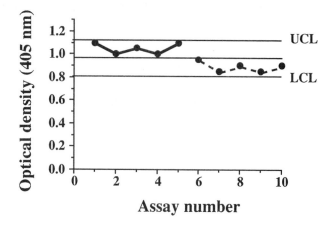

FIG 6.7. A theoretical quality control plot showing upper (UCL) and lower (LCL) confidence limits. The solid and broken lines represent different batches of reagent.

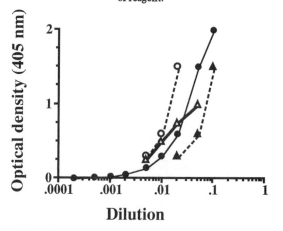

FIG 6.8. Parallelism. The broken lines lie parallel to the reference curve. Dilutions of a third sample (Δ) do not produce a parallel curve.

curves (Fig 6.8). The trouble comes when curves are not parallel. This can be caused by many factors, such as reagent purity, affinity and reagent stability. I have no simple answer to this except to say that if assays are carefully optimised this doesn't happen too often. When it does, always state this in your report/paper and decide whether the variation evident would invalidate your conclusions.

Data handling

Because ELISA is technically easy to perform a considerable amount of data can be generated very quickly. Automated plate-readers make measurements of optical density easy and many can now be linked up to microcomputers so that the results can be processed automatically. I do not wish to support one system rather than another but would strongly recommend that you choose a system that can be shown to do the sort of things you need in your laboratory. For example, we wanted to be able to read the results for several plates from a single standard curve. With a little help from the supplier we were shown how to do this. If you do succeed in automating your data processing I strongly recommend that you continue to look at the raw data carefully to make sure that the assay is performing well. It is all too easy just to accept the values given on the print out.

CHAPTER 7

Assay Optimisation and Trouble Shooting

I would guess that the most important factor that determines the performance of your ELISA is the extent to which it has been optimised. This is also the factor over which you have most control. I am sure that there will be considerable pressure to get results quickly but if you don't set the assay up properly the odds are that you will have to repeat it and it will end up taking longer. That is not to say that you need to reinvent the wheel each time you develop a new assay as some parameters change little. For example, we normally carry out our assays at the same temperature (4°C), with the same incubation times (1 hr), and the same type of plate. But we regularly check new reagents (even different batches of the same reagent), the effect of the pH of the coating buffer, and the concentration of the various detector reagents. It is not a good idea just to use the same conditions as someone else in your department without checking yourself that they are optimal for your assay. It is also worth being bold - so test a wide range of reagent concentrations. The reason for this is that a difference may not be seen at one concentration of the sample but is seen at another. The fact that this can so easily be done makes it inexcusable to do otherwise and the benefits of a properly developed assay will quickly repay the time spent.

Of course, just because one has developed a good assay it does not follow that the results will be useful, only that they are more likely to be correct. However, it is all too frequent that one comes across potentially exciting studies which are marred simply because sloppy technique was used. In this chapter I will go through the various stages of assay optimisation and would thoroughly recommend that new assays and batches of reagents are subjected to appropriate testing.

I have developed my own approach to assay optimisation. I start at the bottom and work up. The bottom in this case is the microtitre plate which serves as an anchor to which all the subsequent reactants must bind. If this fails, which it does remarkably often, then no matter how well the subsequent stages of the assay work, the assay will fail.

CHOICE OF SOLID-PHASE

Table 7.1 lists the things required of a solidphase.

TABLE 7.1.
Requirements of a suitable solid-phase for ELISA.

Requirements of the solid-phase
1 The coating material should actually bind to the plate.
2 Having bound it should not fall off.
3 Having bound firmly it should retain as much of its immunological activity as possible.

Even if all these things are so, it is possible that more subtle, selective changes in the integrity of the coating material will occur. Ways of dealing with these difficulties have been discussed in detail in chapter 4.

PREPARATION OF SOLID-PHASE

Method of testing

The first thing to consider when evaluating the different coating conditions is the method of assessment you will use. It is a waste of time, for example, to test the capability of a microtitre plate coated with a monoclonal antibody to bind a particular antigen if there is no mouse immunoglobulin actually bound to the plate in the first place. So it is important to devise simple experiments to test how well a plate has been coated. This can be established simply by adding an enzyme labelled anti-mouse antiserum. The presence or absence of mouse immunoglobulin bound to the plate can thus be established. Alternatively the binding of different radiolabelled antigens (Fig 7.1) and their subsequent dissociation (see chapter 4) can be tested. I doubt whether we would have learned as much about our assays by simply studying the effect of changing the coating concentration in the assay and I would advise you to test a range of conditions and to study the effect of various parameters using simple and inexpensive methods, before testing the coated plate in the assay.

As well as using enzyme-labelled antibodies and ^{125}I labelled antigens to test the efficiency of coating to these can be used to determine how well IgE was captured by the anti-IgE coated plates using ^{125}I IgE (see table 7.2 below).

Concentration

Probably the most obvious and variable aspect of the coated material will be the concentration used to coat the plate because the capacity of plastic microtitre plates to bind protein is limited. At high concentrations there is a tendency for protein molecules to bind to each other because there is no space

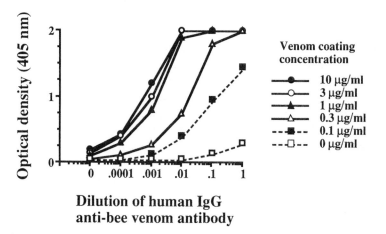

FIG 7.1. Comparison of the binding of different radiolabelled proteins to plastic microtitre plates. The hatched bars are high, and the solid bars low, capacity microtitre plates. PLA2 = bee venom phospholipase A2, BSA = bovine serum albumin, BGG = bovine gamma globulin, β-LACT = β lactoglobulin.

FIG 7.2. Determination of the optimum coating concentration of bee venom antigen for an assay for IgG antibodies to bee venom.

left on the surface of the plastic. Such protein-protein interactions are generally weaker than those between protein and plastic and can result in dissociation of supposedly bound protein during the assay. The range of protein concentrations at which there is no interference with binding to the plastic is called the zone of independent binding. In practice this is typically 1 μg/ml although as much as 10μg/ml can be used without difficulty.

A typical result is shown in figure 7.2. The best concentration to use is the one that gives the steepest dilution curve (the reason for this is discussed in chapter 6). If the coating material is particularly precious it may be necessary to use a suboptimal concentration or to re-use the coating solution in subsequent assays. The precision of the results may of course be affected and this needs to be offset against the saving in source material.

Indeed, there are times when suboptimal conditions can be used to advantage and I know of a case where a friend of mine identified the clones that secreted a particularly high-affinity monoclonal antiserum with a rather insensitive assay. One reason why this particularly high affinity clone was found might be that the immunoglobulin-coated microtitre plates used to screen the clones were only capable of detecting relatively high-affinity antibody. Thus the conditions used should be tailored to the particular needs of the user.

Choice of buffer and pH

The mechanism by which proteins stick to plastic is not completely understood. Certainly charge and hydrophobicity play important roles. The charge expressed by a protein depends on the pH of the buffer in which it is dissolved. By using buffers of different pH it is possible to create optimum conditions for a specific coating material. We routinely test with a pH 9.6 carbonate/bicarbonate buffer, a pH 7.4 phosphate buffered saline and a pH 5.0 acetate/citrate buffer, as described in the appendix. In the example shown in figure 7.3 it can be seen that the pH of the coating buffer has a marked influence on the amount of mouse immunoglobulin that binds to the plate for one monoclonal antibody (Fig 7.3a) while for another (Fig 7.3b) it has no effect. For some antigens, such as wheat gliadin, it is necessary to coat in 70% alcohol. Alcohol is also reported to provide better coating of viral and bacterial extracts. Even distilled water may give enhanced binding to plastic. It is partly a question of trial and error but also of knowing the composition of the coating material. Clearly this is easier for antibodies which only have a single site of activity, their antigen combining site, than for antigens which may have many epitopes. Indeed, under conditions where some antigens are well represented on the surface of the plastic other antigens present in the same preparation may not be.

Time

The time taken to coat a microtitre plate will depend on the concentration of coating solution and the temperature at which this is carried out. Although it is often sufficient to coat for one hour, we usually coat overnight to leave the following day clear for performing the assay. When working with a new preparation it is a good idea to check the effect of coating time.

Blocking vacant protein binding sites on the plate

This is probably the subject which causes more disagreement than any other. We have observed both increased and decreased background activity as a result of blocking. Not all blocking reagents work equally well and the size

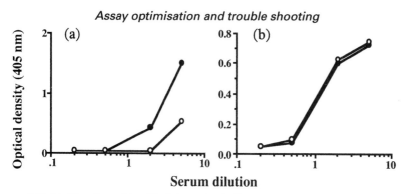

FIG 7.3. The effect of the pH of the coating buffer on the binding of monoclonal antibodies to plastic. Some monoclonal antibodies bind better at a specific pH (a) while others bind equally in different pH coating buffers (b). PBS pH 7.4 = filled circles, bicarbonate buffer pH 9.6 = open circles.

of the protein used to block, for example, may influence this, with small proteins providing a more efficient barrier than large ones.

ASSAY CONDITIONS

Volume

There is relatively little scope for variation here. Microtitre plates can take between 50 and 300 µl of reagent (Table 7.2). Typically we use 100 µl to coat the plate but if greater sensitivity is needed this can be increased, although as little as 50 µl can be used. With less than 50 µl there is a tendency to get uneven coating due to the effects of surface tension.

TABLE 7.2.
The proportion of added antigen (IgE) that binds to anti-IgE coated microtitre plates increases as less is added to the plate. With 50µl of IgE the proportion that binds is greater than with 300µl. 1 IU of IgE = 2.4 ng, NT = not tested.

	IgE added (IU/ml)						
	5	1.5	.5	.15	.05	.015	.005
50µl	45%	40%	75%	80%	77%	86%	88%
300µl	NT	NT	17%	23%	NT	NT	NT

Temperature

At higher temperatures the rate of binding is increased as is the rate of dissociation. For convenience we generally coat at 4 °C overnight or at 37 °C for one hour if a result is needed sooner. I generally carry out all incubations in my assays in the fridge or cold room where the temperature is more controlled.

Molarity & pH

The concentration of buffers used for coating microtitre plates is generally between 0.05M and 0.1M. As long as the particular proteins are stable under these conditions there is no real need to worry. Antibody-antigen reactions normally take place at between pH 6.0 and 9.0. A slight loss of sensitivity may occur at non-physiological pH and it is essential that the same conditions apply to the calibrant as to the sample. A good example of this is assay of proteins in tissue culture supernatants where the pH is 9.0.

Additives

I find it hard at times to see whether I have filled all the wells of a microtitre plate or not. Furthermore, when two reagents are added, as in competitive assays, or when different pH buffers are used, I find it particularly helpful to add coloured dyes. I first saw this when working in a friend's laboratory in Holland. He used a mixture of red and blue dyes to colour the buffers used for RIA. The purpose of this was to show the person doing the assay which stage they were at (there was a subtle shift in pH from 7.4 to 6.0 during the assay). When coating with different buffers I find the same mixture of dyes particularly useful (see appendix). In some cases these can affect the assay, so they should be tested carefully first.

In addition to coloured dyes, two additives are invariably added to the sample buffer to reduce non-specific binding. The first is a detergent. We use one called Tween 20 although I am sure other detergents would do. The concentration used is 0.5% v/v but in our hands a range from 1 to 0.01% has been equally effective. The second additive routinely used is serum. We use 1% horse serum, for reasons of economy, but if there are short circuits (as explained in trouble shooting below) the serum from a specific species must be used; indeed, as a general policy it is a good idea to use a serum from the same species as the enzyme-labelled antibody.

Incubation time

It is important that incubations in ELISA are continued until the rate of binding has slowed down so that the reaction is at or near completion. This is so that any small differences in length of time a particular sample has been incubated will have little or no effect on the amount of colour produced. In practice I have found that for most antigens or antibodies near maximal binding is achieved within 2 hours (Fig 7.4).

SAMPLE

Just as there is a limited range of protein concentrations that bind effectively to plastic, there is a limit to the amount of sample that can be measured. In practice this is rather similar between assays over a concentration range of 1-1000 ng/ml. Very low concentrations require special modifications (see chapter 5) and higher concentrations need dilution. There are other reasons

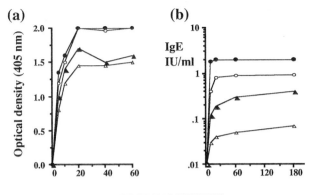

TIME IN MINUTES

FIG 7.4. Optimum incubation time for (a) monoclonal anti- IgG1 (filled circles), IgG2 (open circles), IgG3 (filled triangles) and IgG4 (open triangles); (b) for binding of IgE to anti-IgE coated microtitre plates.

why the concentration of sample should not be too high. At high serum concentrations there is a greater tendency for low-affinity antibodies to bind. In addition, at high immunoglobulin concentrations a small but significant proportion will bind non-specifically to the coating antigen or directly to the plate. In practice it is rarely practical to assay serum samples at a dilution of less than 1/50.

LABELLED DETECTOR

Considerations of diluent, temperature and length of incubation for the detector are the same as for the sample. Above all the concentration of detector needs to be optimal. The suggested working dilution provided by suppliers of enzyme-labelled reagents should always be checked in your assay. It is not possible for the manufacturer to know what you need and it is no good saying "It was on the bottle".

Competitive assays

The simplest form of detector is antibody that is directly conjugated to the enzyme. The main drawbacks with such reagents is the work involved in isolating the antibody concerned and labelling it. Those who are inexperienced would be well advised to use commercially prepared reagents first. The detector can be an antibody or an antigen attached directly or indirectly to an enzyme. When the detector is used in a competitive fashion (chapter 3, figs 3.4 & 3.5) it should be used at a limiting concentration. This can be determined (Fig 7.5) by adding sufficient detector, in the absence of sample, to give an optical density as close as possible to the maximum that can be read (2-2.5). This will normally be in the steep part of the binding curve and will give the widest detectable range practical.

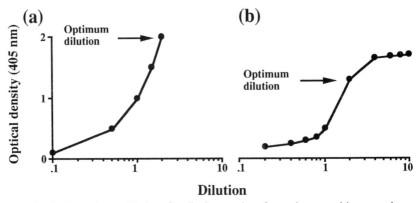

FIG 7.5. The optimum dilution of antibody or antigen for use in competitive assays is determined by titration. If there is no plateau (a) then a dilution close to the highest recorded OD reading can be used. If the curve is sigmoid (b) then the top of the steep part of the curve should be used.

Non-competitive assays

The concentration of detector needed must be determined by experiment. As for coating the plates a concentration should be chosen such that small errors in diluting do not adversely effect the assay. For non-competitive assays (chapter 3 figs 3.1-3.3) the detector should be in excess. If there are a number of reagents used in the detection stage of the assay (e.g. monoclonal antibody followed by enzyme labelled anti-mouse antibody), it is essential to optimise the first layer (Fig 7.6a) and then the subsequent stages (Fig 7.6b). The use of particular antibody fragments and labelling procedures are discussed in detail in chapter 6.

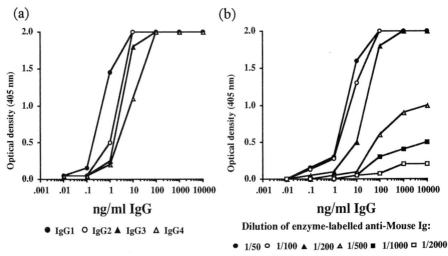

FIG 7.6. Optimal concentrations of (a) monoclonal anti-IgG and (b) enzyme-labelled anti-mouse Ig.

SUBSTRATE

For the enzyme-substrate reaction, the conditions of pH, molarity and temperature should be optimal. If, however, too much colour is generated, we have sometimes found that some information can be obtained by diluting an aliquot from each well. However, if the rate of colour development is wildly different from predicted it usually means that something in the assay has changed and that conditions are no longer optimal.

TROUBLE SHOOTING

Clearly, the best way of dealing with problems is to ensure that the assay is set up properly the first place and monitor the activity of the reagents used. We have found that antigen preparations used to coat the plate can change with time. We periodically re-titrate these to ensure that we are using optimal conditions. Edge effects can be seen and are usually the result of a poorly optimised assay. Wherever possible make assays so robust that they will tolerate a very ham-fisted operator.

To find out why an assay has gone wrong it is essential to isolate the problem. There are so many variables in ELISA that it is usually impossible to say why an assay has failed without further investigation. This means breaking the assay down into its component parts. Of the things that can go wrong I have a hierarchy which goes something like this:

Most trouble			Least trouble
Coating	> Detector	> Enzyme/substrate	> Sample

Assay design

It may seem obvious, but the appropriate assay design must be used. It is surprising how often simple flaws at this stage of planning give rise to difficulties later. Certainly the more complex the protocol, the more likely it is that errors will occur. Some types of assay, for example the two-site assay, give less trouble compared with, say, competitive assays. You have to weigh up the ease with which a particular assay can be set up against the precision you will need. Before trying out a new assay design I show it to the others in my lab. It is much easier if flaws in the design are spotted at this stage.

Short circuits

A short circuit in an immunoassay occurs when the reagents bind to each other as well as, or instead of, the sample. It is distinguished from non-specific binding in being a specific immune reaction. The likelihood of short circuits increases in proportion to the number of reagents used. It is generally a good idea to include control wells which contain every combination of reagents except the sample. As an example, I have used an assay for human IgG subclass proteins (Fig 7.7) in which the alkaline phosphatase-labelled rabbit anti-human IgG bound to the mouse anti-human IgG subclass antibody on the plate. In this case the solution was to absorb the mouse antibody against normal mouse serum agarose and to include 1% normal mouse serum in the conjugate diluent.

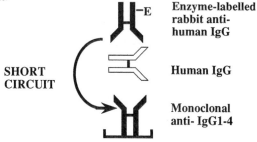

SHORT CIRCUIT

Enzyme-labelled rabbit anti-human IgG

Human IgG

Monoclonal anti- IgG1-4

FIG 7.7. Short circuits most commonly occur when one antiglobulin antibody cross reacts with another.

Cross-reactions

Assays are only as good as their component parts. If the immune reagents used cross-react, then they won't work very well. Such problems can be overcome by pretreating, absorbing or adding inhibitors to the assay, as described below. Not all cross reactions can be absorbed out and under these circumstances it is best to look for an alternative reagent.

PLATES

Batch-to-batch variation

Difficulties with plates are mercifully less common than they used to be. In the future I hope that plates specifically designed to bind proteins will become available. Nevertheless it is a good idea to buy several boxes at a time and to test new batches in parallel with the old plates when they come in. I generally use one of our most robust and one of the most finicky assays to evaluate them. If you do find that the new batch works less well, complain; if they work better, also tell the manufacturer so that he or she is able to continue to improve the product.

Blocking

The most common complaint is that background binding is too high. A common method of dealing with this is to add 'a load more' protein. In my experience this often creates more problems, as plates on their own do not bind very much immunoglobulin in a buffer containing a little Tween 20 and added serum. If you try you will find that as soon as you start blocking with protein the plate often becomes more sticky. It is really a question of trial and error.

Edge effects

Another common complaint is of edge effects. These are manifested by greater colour development in the outer wells. They are caused by differences in temperature - these wells generally run slightly hotter. This can be reduced by refraining from stacking the plates directly on top of each other. More importantly, if the coating conditions are arranged so that the appropriate reagents are in excess, such problems are usually avoided.

REAGENTS

Batch-to-batch variation

There is nothing worse than running a large assay which flops and then finding that it is due to a batch of a new reagent which is inactive or is less potent than that used previously. People usually test those that they make themselves but often forget to do this with new batches of a commercial reagent. It need not take up much time to test these and you will certainly save yourself a lot of problems later.

Pretreatment

It is not uncommon for ELISA users to expect more of their reagents than was intended. For example, is it really reasonable to expect a commercial rabbit anti-mouse immunoglobulin antibody not to bind to rat immunoglobulin? In such cases it will be necessary to process the reagent, prior to use, by absorption.

Diluent

A lot of problems can be avoided if the diluent contains serum from the same species as those reagents used in the test. Most animal sera are freely available commercially. The detector can be pre-incubated with this to remove cross-reacting antibodies prior to use. For example, antibody to human IgG which might react with a rabbit IgG which was bound to the plastic could be pre-incubated for an hour with 1% normal rabbit serum. Alternatively, to assess binding of human IgE antibodies to allergens of the dog other than dog serum albumin, an excess of dog serum albumin can be used to pretreat the human serum. These are not necessarily ideal solutions but on occasion they can prove very useful.

Absorption

The best method of removing cross-reacting antibodies is to absorb the antibody with antigen-coated agarose. This means that the problem has been removed. Absorption with antibody-coated agarose (nowadays called negative affinity chromatography) is also a highly effective method of removing contaminants from antigen preparations.

CHAPTER 8

Future Developments

In this chapter I would like to look at the future of ELISA and its application. Probably the most important single virtue of ELISA is its ease of use: the widespread availability of enzyme-labelled reagents and suitable chromogenic substrates and the convenience of microtitre plates. The greater safety of ELISA too has helped to popularise the technique and the tremendous amplification of the signal that enzymes achieve will ensure that it remains at the forefront of immunoassay in terms of sensitivity.

I believe that ELISA will be used in three different ways. First in research laboratories for the new substances that are produced (eg using molecular biological techniques). Second, as on-site or home tests which can be carried out by the inexperienced. The arrival of my first child, 5 years ago, was heralded by an agglutination test, whereas the birth of the second, 10 months ago, was predicted by a dip stick ELISA. Finally, I believe ELISA will become the method used in automated immunoassay systems.

TABLE 8.1.
Future development of ELISA

The future of ELISA

1   Sensitivity
2   Speed
3   Convenience

SENSITIVITY

Although ELISA is a very sensitive technique, there is a need for ever more sensitive methods. Bioassays for interleukins and other cell products which are locally active, for example, are much more sensitive than ELISA. More than any other in recent years Professor Ishikawa's group in Japan have investigated the factors that govern sensitivity. The limiting factor seems to be the form of the detector.

It doesn't appear to be the affinity of the immune response that limits sensitivity, for once an antigen or antibody is bound to a solid-phase its behaviour deviates from the law of mass action, as discussed at the end of chapter 2. The law of mass action makes the assumption that the movement of molecules in solution is not affected by the reaction vessel. This is almost certainly true for reactions wholly carried out in the fluid-phase. When one of the reactant is bound to the solid-phase something else, I believe, is happening. The antibody that dissociates is much closer to other antigen molecules than it would be if both were in solution. In other words, concentration at the surface of the solid-phase is almost infinite and with a monoclonal antibody to human IgE we can detect as little as 10 pg/ml of IgE (see fig 5.8 chapter 5). The monoclonal antibody has an affinity of 10^{-11}M but the concentration of IgE is about 10^{-13}M. I believe that Ishikawa and his staff can now detect 1000 times less than this! Sensitive combination techniques utilising enzyme labels and radioactive substrates which are very sensitive have also been described but are nowhere near as convenient and will not I believe be widely used.

SPEED

There is little doubt in my mind that the demand for ever faster assays is unremitting. Not just to increase the turn around time for routine samples or to help make it possible to reach decisions about the selection of clones but I believe that greater speed will enable ELISA to be brought into the places where it is needed (see below). In this section I will consider the factors that govern the speed of ELISA.

Concentration

There is no doubt that if the concentration of one or more of the components is increased the speed of the reaction will be accelerated. However, there are limits to how much of a particular component is used or to how high the concentration can go without other factors being affected.

Reactive area

Most of the antibody or antigen in an assay is some distance from the solid-phase. It takes a quite a long time before all of the molecules in the fluid-phase have an opportunity to bind. If a very small volume of sample were introduced into a reaction vessel with a very high surface area, then the reaction should be almost instant. Using a highly efficient wash through device we have been able to reduce the time taken for an assay for IgE and IgE antibody from 3 days to 1 hour simply by reducing the volume of the reagents so that the majority of the antibody in the sample was in close contact with the antigen-coated fibres of the paper disc (Fig 8.1). Because the opportunity for antibody to bind the antigen was greater in the micro assay, the rate of binding was increased substantially (Fig 8.2).

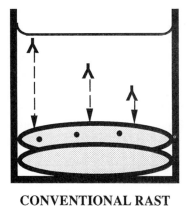

CONVENTIONAL RAST   **MICRO-RAST**

FIG 8.1. Close contact between all of the sample and the coated carrier can be accomplished by a reduction in the volume of sample used in the micro RAST.

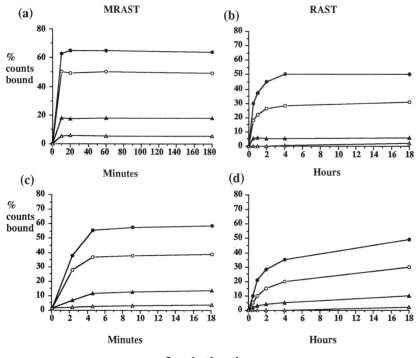

Incubation time

FIG 8.2. The increased rate of binding of IgE antibody (a) and labelled anti-IgE (b) in the micro RAST as compared with the conventional assay (c & d). Closed circles = neat serum, open circles = 1/10, closed triangles = 1/100 and open triangles = 1/1000.

The rate of dissociation was unchanged, however, so that the equilibrium was shifted in favour of AbAg (Fig 8.3). Others too have noted that close contact between antigen and antibody speeds up the reaction and recently a commercial company has introduced an apparatus for pulling the reagents through nitrocellulose membranes.

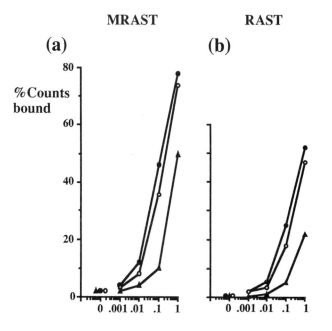

Dilution of serum pool

FIG 8.3. The closer contact between the detector antibody and its target in the microassay (a) compared with the conventional assay (b) resulted in an increase in sensitivity with different amounts of sample($\bullet$ =300µl)($\bigcirc$ =50µl)($\blacktriangle$ =5µl).

Signal

Another limiting factor is the enzyme substrate. With an ingenious amplification system (see chapter 6) developed by Professor Self, Newcastle, it is possible to increase sensitivity to a lower detection limit of 3000 enzyme molecules. Fluorogenic substrates too offer tremendous sensitivity but whatever method of amplification is used the background will also go up and an increase in sensitivity will only be achieved if allied to a cleaner assay system.

CONVENIENCE

The fact that ELISA is relatively hazard free, has a visible end point and is easy to performance make it suitable to take out of the laboratory where an increasing proportion of diagnostic tests are being carried out. These tests are generally non-invasive, simple and easy to use. They should employ as few steps as possible and should not require precise pipetting of the sample. The sample should be a readily available body fluid or tissue and the result should be easy to interpret. To facilitate this such tests should incorporate internal controls and should have a simple readout such as the ICONTM system where the result is seen as a + or -.

The assay should run to completion so that the timing of different steps is not critical. These tests are, in many ways analogous to the old fashioned litmus paper test for acids and alkalis which is still in use today or the alcohol breath test which has been of so much to help in combating drunken driving. These ex-laboratory tests may be performed in doctors' clinics (clinic/office tests) or by the patient (self/home tests). They may also be performed in the workplace (on-site tests) such as on farms or in factories.

Clinic/office tests

As their name implies these tests are carried out at the doctors' surgery. They have the advantage of being able to provide a result when the patient is still there or shortly afterwards. This is especially useful for identifying the relevant drug in cases of overdose or the appropriate species of snake in snake bite victims. They are usually carried out by the nurse or another clinical assistant.

Self/home tests

The use of self or home tests for non-life threatening conditions is rapidly increasing. The most common test materials are urine, blood and faeces. Although other body fluids such as saliva, milk, seminal fluid and vaginal secretions could also be tested. The requirements are much the same as for clinic/office based tests although in general it is preferable that collection of the sample is non-invasive. The main current and future uses of these tests are listed below:

TABLE 8.2.
Applications for ex-laboratory tests

Current and future uses of self/home tests

Current uses	Future uses
Blood glucose levels.	Therapeutic drug monitoring.
Prediction of ovulation.	Diagnosis of infectious agents.
Pregnancy tests.	Diagnosis of sexually transmitted diseases.
	Detection of breast or cervical cancer.
	Diagnosis of cardiovascular problems.
	In vivo monitoring.

I am, however, concerned about the dangers of the indiscriminate use of such tests which could be a waste of precious resources. Many people keep thermometers in their house to confirm a fever. The clinical information gained is of little practical value but the cost is low and no harm is done. Self or home tests will cost much more and besides wasting money they may increase patient anxiety.

On-site tests

These could be carried out in the workplace - be it a food processing plant, a farm, or at immigration control. They offer the same advantages as clinic tests in providing an answer where it is needed. In the field of veterinary medicine they can be used to monitor fertility and to detect infectious disease. They may also be used to detect toxins and antibiotic residues in animal feeds and animal products.

THE TECHNOLOGY

Dip-sticks

These were probably the earliest potential ex-laboratory tests involving enzyme immunoassay. The dipsticks were made of plastic of a higher capacity matrix such as nitrocellulose or CNBr-activated paper bonded to a plastic stick (Fig 8.4). They were limited by the capacity of the plastic strip or by the diffusion of the samples and reagents through the matrix.

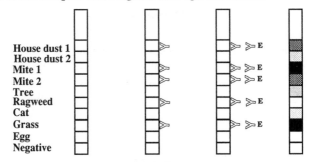

FIG 8.4. The multiallergen dip stick after Novey et al 1987.

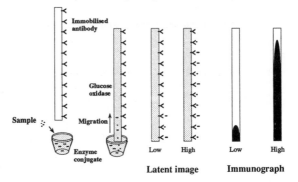

FIG 8.5. The test strip immunoassay (after Litman et al 1983).

Immunofiltration

A number of different systems have been developed in which the porous nature of the solid-phase matrix is used to separate or concentrate the sample. In the test strip assay antibody to morphine is bound to a strip of paper (Fig 8.5). The blood sample is mixed with enzyme-labelled morphine and the test strip dipped into it. As the sample enters the test strip blood cells bind to the plant lectin concanavalin A, which has been bound to the bottom portion of the strip. The unlabelled morphine present in the blood sample competes with the enzyme-labelled morphine for anti-morphine antibody thus forcing it to travel higher up the strip in search or free antibody.

Another approach is to use concentration through a porous membrane as in the immunoconcentration ICON™ assay in which (Figure 8.6) the antigen is captured by filtration through a membrane filter to which an antibody has been bound. The antigen thus bound to the membrane is visualised by the addition of enzyme-labelled antibody followed by a chromogenic substrate with an insoluble product.

Self-contained immunoassay

In the self-contained immunoassay all the reactants are kept together in a single reaction vessel. A good example of this is the Biomat-EIA™ in which the reagents are sequestered in immiscible layers which also serve to separate the bound and unbound reactants (Figure 8.7). In the upper layer are antibody coated beads premixed with an enzyme conjugate. The sample is added to the vessel via a self-sealing septum and after incubation, the tube is centrifuged forcing the beads through a water immiscible layer to separate them from the unbound conjugate. The beads enter the bottom layer which contains the developing solution and the result can be read visually in a specially designed colorimeter. A similar approach is used in the Sucrosep™ technique in which a relatively dense sucrose solution is used to separate the free and bound fractions.

AUTOMATION

There seems little doubt that we are moving toward an increasingly automated laboratory environment. A good example of such a system is the radial partition assay (Fig 8.8). Antigen or antibody is bound to the centre of a glass fibre tab followed in sequence by the other components of the assay. Washing is effected by spinning the tab and each successive reagent displaces its predecessor. The signal (fluorescence) is read from the centre of the tab. The assay can detect nanogram quantities of Thyroxine or ferritin in 7 minutes. There is no doubt that recent advances in the area of robotics has made this much more feasible. Assays will need to be specifically tailored to make use of this form of apparatus. Those where all the reagents can easily be moved around the system (fluids) will be favoured. Some assays will measure changes in electrical conductivity as the substrate is digested by the enzyme.

(a)

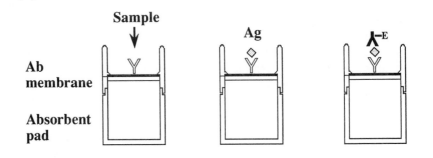

Ab membrane

Absorbent pad

(b)

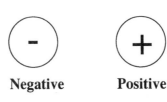

FIG 8.6. The immunoconcentration assay ICON™. (a) The concentration device and assay steps. (b) An end-view.

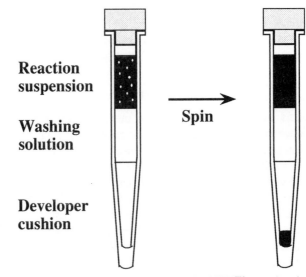

Reaction suspension

Washing solution

Developer cushion

Spin

FIG 8.7. The self contained immunoassay. In the BioMAT™ assay immiscible layers sequester the reagents and wash them allowing the whole assay to be self contained.

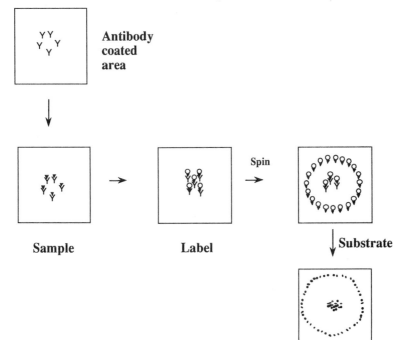

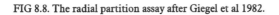

FIG 8.8. The radial partition assay after Giegel et al 1982.

Bibliography

BOOKS

Collins, W.P. (1985). Alternative immunoassays. John Wiley and Sons, Chichester, UK.

Collins, W.P. (1985). Complimentary immunoassays. John Wiley and Sons, Chichester, UK.

Ishikawa, E., Kawia, T., Miyai, K. (1981). Enzyme immunoassay. Igaku-Shoin, Tokyo, Japan.

Kemeny, D.M. and Lessof, M.H. (1983). Recent developments in the RAST and other solid-phase immunoassays. Experta Medica 1983, Amsterdam.

Kemeny DM and Challacombe SJ. (1988) ELISA and other solid-phase immunoassays. Theoretical and practical aspects. John Wiley & Sons, Chichester, UK.

Maggio, T. (1979). The enzyme immunoassay. CRC press, NY, USA.

Ngo, T.T. and Lenhoff, H.M. (1985). Enzyme-mediated immunoassay. Plenum Press, NY, USA.

Tijssen, P. (1985). Practice and theory of enzyme immunoassays. Vol 15 in Laboratory techniques in biochemistry and molecular biology. Eds Burdon RH & Knippenberg. Elsevier, Amsterdam, Holland.

Voller, A., Bidwell, D.E., and Bartlett, A. (1979). The enzyme-linked immunosorbent assay (ELISA). Dynatech Europe, UK.

Voller A, Bartlett and Bidwell (1981). Immunoassays for the 80s. MTP press, Lancaster, UK. pp. 457-479.

REVIEWS

Avrameas, S., Ternynck, T., and Guesdon, J-L. (1978). Coupling of enzymes to antibodies and antigens. Scand J Immunol 8 (Suppl 7) : 7-23.

Blake, C., and Gould, B.J. (1984). Use of enzymes in immunoassay techniques. A review. Analyst. 109: 533.

Guilbault, G.G. (1968). Use of enzymes in analytical chemistry. Anal Chem, 40: 459.

Kemeny, D.M., and Challacombe, S.J. (1986). Advances in ELISA and other solid-phase immunoassays. Immunology Today, 7: 67.

A practical guide to ELISA

Kemeny, D.M. (1987). Immunoglobulin and antibody assays. In: Allergy an international textbook.Eds Lessof, M.H., Lee, T.H., and Kemeny, D.M. John Wiley and Sons Ltd, Chichester, UK; 319.

Kemeny, D.M., and Chantler, S. (1988) An introduction to ELISA. ELISA and other solid-phase immunoassays. Theoretical and practical aspects. Eds Kemeny DM, Challacombe SJ. John Wiley and Sons Ltd, Chichester, UK; 1.

Kemeny, D.M., and Challacombe, S.J. (1988). Microtitre plates and other solid-phase supports. In: ELISA and other solid-phase immunoassays. Theoretical and practical aspects. Eds: Kemeny DM, Challacombe SJ John Wiley and sons Ltd, Chichester, UK; 30.

Kemeny, D.M. (1988). The Modified sandwich ELISA (SELISA) for detection of IgE and other antibody isotypes. In: ELISA and other solid-phase immunoassays. Technical and theoretical aspects. Eds Kemeny DM, Challacombe SJ. John Wiley & sons Ltd, Chichester, UK; 197.

Kemeny, D.M. (1990). Immunological tests in allergy. Current Opinions in Immunology (In press).

Landon, J. (1977). Enzyme-immunoassay: techniques and uses. Nature, 268: 483.

Van Weemen, B.K. (1985). ELISA: Highlights of the present state of the art. J. Virol Methods, 10: 371.

Yolken, R.H (1982). Enzyme immunoassays for the detection of infectious antigens in body fluids: current limitations and future prospects. Rev Infect Dis, 4: 35.

ORIGINAL ARTICLES

Aalberse, R.C., Van Zoonen, M., Clemens, J.G.J., and Winkel, I. (1986). The use of hapten-modified antigens instead of solid-phase coupled antigens in a RAST-type assay. J Imm Methods, 87: 51.

Avrameas, S. (1969). Coupling of enzymes to proteins with glutaraldehyde. Use of conjugates for the detection of antigens and antibodies. Immunochemistry, 5: 43.

Avrameas, S., and Ternynck, T. (1971). Peroxidase labelled antibody and Fab conjugates with enhanced intracellular penetration. Immunochemistry, 8: 1175.
Axen, R., Porath, J., and Ernback, S. (1967). Chemical coupling of peptides and proteins to polysaccharides by means of cyanogen halides. Nature, 214: 1302.

Berkowitz, D.M., and Webert, D.W. (1981). The inactivation of horseradish peroxidase by a polystyrene surface. J. Imm Methods, 47: 121.

Boorsma, D.M., and Kalsbeek, G.L. (1975). A comparative study of horseradish peroxidase conjugates prepared with a one or two step method. J Histochem. Cytochem, 23: 200.

Boorsma, D.M., and Steefkerk, J.G. (1976). Peroxidase-conjugate chromatography isolation of conjugates prepared with glutaraldehyde or periodate using polyacylamide-agarose gel. J. Histochem Cytochem, 24: 481.

Butler, J.E., McGiven, P.L., and Swanson, P. (1978). The amplified ELISA: principles and applications for antibodies and the distribution of antibodies and antigens in biochemical studies. J Imm Methods, 20: 365.

Cantarero, L.A., Butler, J.E., and Osborne, J.W. (1980). The adsorptive characteristics of proteins for polystyrene and their significance in solid-phase immunoassays. Analytical Biochemistry, 105: 375.

References

Catt, K., and Tregear, G.W. (1967). Solid-phase radioimmunoassay in antibody-coated tubes. Science, 158: 1570.

Ceska, M., and Lundkvist, U. (1972). A new and simple radioimmunoassay method for detection of IgE. Immunochemistry, 9: 1021.

Chen, R., Weng, L., Sizto, N.C., Osonio, B., Hsu, C.J., Rodjers, R., and Litman, D.J. (1984). Ultrasound accelerated immunoassay as exemplified by enzyme immunoassay of choriogonadotrophin. Clin Chem, 30: 1446.

Challacombe, S.J., Biggerstaff, M., Greenall, C., and Kemeny, D.M. (1986). ELISA detection of human IgG subclass antibodies to Streptococcus mutans. J Imm Methods. 87: 95.

Ciclitira, P.J., Ellis, H.J., Richards, D., and Kemeny, D.M. (1986). Gliadin-specific IgG subclass antibodies in patients with coeliac disease. Int Archs All Appl Immun, 80: 258.

Cleveland, P.H. Richman, D.D., Redfield, D.C., Disharon, D.R., Binder, P.S. and Oxman, M.N. (1982). Enzyme immunofiltration technique for rapid viral diagnosis of herpes simplex virus eye infections in a rabbit model. J. Clin. Microbiol, 16: 675.

Devey, M.E., Lee, S.D., Richards, D., Kemeny, D.M. (1989). Serial studies on the functional affinity and heterogeneity of antibodies of different IgG subclass to phospholipase A2 in response to bee venom immunotherapy. J All Clin Immun 84: 326.

Engvall, E., and Perlmann, P. (1971). Enzyme linked immunosorbent assay (ELISA): quantitative assay of IgG. Immunochemistry, 8: 871.

Freytag, J.W., Dickinson, J.C., and Tseng, S.Y. (1984). A high sensitivity affinity-column-mediated immunometric assay, as exemplified by digoxin. Clin Chem, 30: 417.

Guesdon, J-L., Ternynck, T., and Avrameas, S. (1979). The use of avidin-biotin interaction in immunoenzymatic techniques. J Histochem. Cytochem, 27: 1131.

Giegel, Y.L., Brotherton, M.M., and Cronin, P., et al. (1982). Radial partition immunoassay. Clin Chem, 28: 1894.

Hargreaves, W.R. and Harris, G.H. (1986). Self-washing heterogeneous immunoassay system applied to detection of a cancer-related peptide. Clin Chem, 32: 1323-1327.

Hamilton, R.G., and Adkinson, N.F. (1985). Naturally occurring carbohydrate antibodies: Interference in solid-phase immunoassays. J Imm Methods, 77: 95.

Hancock, K., and Tsang, U.C.W. (1986). Development of a FAST-ELISA for detecting antibodies to Schistosoma Mansoni. J Imm Methods, 93: 89.

Hendrey, R.M., and Herremann, J.E. (1980). Immobilisation of antibodies on nylon for use in enzyme-linked immunoassays. J Imm Methods, 35: 285.

Herremann, J.E., and Collins, M.F. (1976). Quantitation of immunoglobulin absorption to plastics. J Imm Methods, 10: 363.

Howell, E.E., Nasser, J., Schray, K.J. (1981). Coated tube enzyme immunoassay: Factors affecting sensitivity and effects of reversible protein binding to polystyrene. J Immunoassay. 2: 205.

Holmes, N.J., and Parnham, P (1983). Enhancement of monoclonal antibodies against HLA-A2 is due to antibody bivalency. J Biol Chem, 258: 1580.

Ishikawa, E., Imagawa, M., Yoshitake, S., Niitsu, Y., Urushizaki, I., Inada, N., Kanazawa, R., Tachibana, S., Nazakawa, N., and Ogawa, H. (1982). Major factors limiting sensitivity of sandwich enzyme immunoassay for ferritin, immunolgobulin E and thyroid stimulating hormone. Ann Clin Biochem, 19: 379.

Johannsson, A., Ellis, D.H., Bates, D.L., Plumb, A.M., and Stanley, C.J. (1986). Enzyme amplification for immunoassays. Detection limit of one hundreth of an attomole. J Imm Methods, 87: 7.

Kemeny, D.M., Lessof, M.H., and Trull, A.K. (1980). IgE and IgG antibodies to bee venom measured by a modification of the RAST. Clin Allergy, 10: 413.

Kemeny, D.M., Frankland, A.W., Fahkri, Z.I., and Trull, A.K. (1981). Allergy to castor bean in the Sudan: measurement of serum IgE and specific IgE antibodies. Clin Allergy, 11: 463.

Kemeny, D.M., and West, F.B. (1981). An improved method for washing paper discs with a constant flow washing device. J Imm Methods, 49: 89.

Kemeny, D.M., Harries, M.G., Youlten, L.J.F., Mackenzie-Mills, M., Lessof, M.H. (1983). Antibodies to purified bee venom proteins and peptides. I. Development of a highly specific RAST for bee venom antigens and its application to bee sting allergy. J All Clin Immun, 71: 505.

Kemeny, D.M., Urbanek, R., Samuel, D., and Richards, D. (1985). Improved sensitivity and specificity of sandwich, competitive and capture enzyme-linked immunosorbent assays for allergen specific antibodies. Int Arch All Appl Immun, 77: 198.

Kemeny, D.M., Urbanek, R., Samuel, D., and Richards., D. (1985). Increased sensitivity and specificity of a sandwich ELISA for measurement of IgE antibodies. J Imm Methods, 78: 212.

Kemeny, D.M., Urbanek, R., Samuel, D., Richards, D., and Maasch, H. (1986). The use of monoclonal and polyspecific antisera in the IgE ELISA. J Imm Methods 87: 45.

Kemeny, D.M., Urbanek, R., Richards, D., and Greenall, C. (1987). The development of a semi-quantitative enzyme-linked immunosorbent assay for quantitation of IgG subclass antibodies. J Imm Methods, 96: 47.

Kemeny, D.M., and Lessof, M.H. (1987). The immune response to bee venom. II. Quantitation of the absolute amounts of IgE and IgG antibody by saturation analysis. Int Archs All Appl Immun, 83:113.

Kemeny, D.M., and Richards, D. (1987). ELISA for the detection of total IgE: Speed and sensitivity. Immunological techniques in microbiology 24: 47.

Kemeny, D.M., and Richards, D. (1988). Increased speed and sensitivity of a micro radioallergosorbent test (MRAST). J Imm Methods, 108: 105.

Kemeny, D.M., Richards, D., Johannsson, A. and Durnin, S. (1989). Ultrasensitive enzyme-linked immunosorbent assay (ELISA) for the detection of picogram quantities of IgE. J Imm Methods, 120: 251-258.

Leaback, D.H., and Creme, S (1980). A new experimental approach to fluorimetric enzyme assays employing disposable micro-reaction chambers. Analyt Biochem, 106: 314.

Lehtonen, O.P., and Viljaken, M.K. (1980). Antigen attachment in ELISA. J Imm Methods, 34: 61.

Litman, D.J., Lee, R.H., Jeong, H.J., Tom, H.K., Stiso, S.N., Sizto, C.C. and Ullmann, E.F. (1983). An internally referenced test strip immunoassay for morphine. Clin Chem. 29: 1598.

References

Macy, E., Kemeny, D.M., and Saxon, A. (1988). Enhanced ELISA: How to measure less than 10 picograms of a specific protein (immunoglobulin) in less than 8 hours. FASEB Journal, 2: 3003.

Miles, L.E.M., and Hales, C.N. (1968). Labelled antibodies and immunological assay systems. Nature, 219: 186.

Nakane, P., and Pierce, G.B. (1966). Enzyme-labelled antibodies. Preparation and application to localization of antigens. J Histochem. Cytochem, 14: 92.

Nakane, P.K., and Kawaoi, A. (1974). Peroxidase-labelled antibody. A new method of conjugation. J. Histochem. Cytochem, 22: 1084.

Novey, H.S., Immam, A.A., Orgel, H.A. et al (1987) Multi allergen dipstick screening test for specific serum IgE. J All Clin Immunol, 79: 235.

Oliver, D.G., et al (1981). Thermal gradients in microtitration plates. Effects on enzyme-linked immunoassay. J Imm Methods, 42: 195.

Oreskes, I., and Singer, J.M. (1961). The mechanism of particulate carrier reactions. I. Adsorption of human globulin to polystyrene latex particles. J Immunol, 86: 338.

Pachmann, K., and Leibold, W. (1976). Insolubilisation of protein antigens on polyacrylamide plastic beads using poly-l- lysine. J Imm Methods, 12: 81.

Park, H.A. (1978). A new receptacle for solid-phase immunoassays. J Imm Methods, 20: 349.

Pesce, A.J., Ford, D.J., Gaizutis, M., and Pollak, V.E. (1977). Binding of proteins to polystyrene in solid-phase immunoassays. Biohem Biophys Acta, 492: 399.

Phillips, D.J., Reimer, C.B., Wells, T.W., and Black, C.M. (1980). Quantitative characterisation of specificity and potency of conjugated antibody with solid-phase, antigen bead standards. J Imm Methods, 34: 315.

Place, J.D., and Schroeder, H.R. (1982). The fixation of anti- HBs Ag on plastic surfaces. J Imm Methods, 48: 251.

Porstmann, B., Porstmann, T., Nugel, E., and Evers, U. (1985). Which of the commonly used marker enzymes gives the best results in colorimetric and fluorimetric immunoassays: horseradish peroxidase, alkaline phosphatase or B-galactosidase. J Imm Methods, 79: 27.

Rotmans, J.P., and Scheven, B.A.A (1984). The effect of antigen cross-linking on the sensitivity of the enzyme-linked immunosorbent assay. J Imm Methods. 70: 53.

Rubin, R.L., Hardtke, M.A., and Carr, R.I. (1980). The effect of high antigen density on solid-phase radioimmunoassays for antibody regardless of immunoglobulin class. J Imm Methods, 33: 287.

Salonen, E.M., and Vaheri, A. (1979). Immobilisation of viral and mycoplasmal antigens and of immunoglobulins on polystyrene surface for immunoassays. J Imm Methods, 30: 209.
Schuurs, A.H., and Van Weemen, B.K. (1977). Enzyme immunoassays. Clin Chim Acta, 81: 1.

Self, C.H. (1985). Enzyme amplification - a general method applied to provide an immune assisted assay for placental alkaline phosphatase. J Imm Methods, 83: 89.

Sharpe, S.L., Cooreman, W.M., Bloome, W.J., and Lackman, G.M. (1976). Quantitative enzyme immunoassay - current status. Clin Chem, 22: 733.

Thorpe, S.C., Kemeny, D.M., Panzani, R., and Lessof, M.H. (1987). The relationship between total serum IgE and specific IgE in castor bean sensitive patients from Marseille. Int Archs All Appl Immun, 82: 456-460.

Ternynck, T., and Avrameas, S. (1977). Conjugation of p- Benzoquinone treated enzymes with antibodies and Fab fragments. Immunochemistry, 14: 767.

Underdown BJ, James O, and Knight A. (1983). Development of ultrasensitive fluorimetric enzyme immunoassay for IgE. In: Recent developments in RAST and other solid-phase immunoassay systems. Eds: Kemeny, D.M., and Lessof, M.H. Excerptia Medica, Amsterdam; 67.

Urbanek, R., Kemeny, D.M., and Samuel, D. (1985). Use of the enzyme-linked immunosorbent assay for measurement of allergen-specific antibodies. J Imm Methods, 79: 123.

Urbanek, R., Kemeny, D.M., and Richards, D. (1986). The subclass of IgG anti-bee venom antibody produced during bee venom immune therapy and its relationship to long-term protection from bee stings and following termination of venom immunotherapy. Clin Allergy, 16: 317.

Van Weemen, B.K., Schuurs, A.H.W.M (1971). Immunoassay using antigen-enzyme conjugates. FEBS Lett, 15: 232.

Voller, A., Bidwell, D.E., Huldt, G., and Engvall, E (1974). A microplate method of enzyme linked immunosorbent assay and its application to malaria. Bull Wld Hlth Org, 51: 209.

Valkirs, G.E. and Barton, R. (1985) Immunoconcentration TM - A new format for solid-phase immunoassays. Clin Chem. 31: 1427-1431.

Wright, J.F. and Hunter, W.M. (1983). The sucrose layering separation: a non-centrifugation system. In: Immunoassays for Clinical Chemistry. Eds. Hunter, W.M. and Corrie J.E.T. Churchill Livingstone, Edinbrugh; 170.

Zuk, R.F., Ginsberg, V.K., Houts, T., Rabbie, J., Merrick, H., Ullman, E.F., Fischer, M.M., Sizto, C.C., Stiso, S.N. and Litman, D.J. (1985). Enzyme immunochromatography - A quantitative immunoassay requiring no instrumentation. Clin. Chem, 31: 1144-1150.

Appendix of Methods

APPARATUS

Of all laboratory immunological techniques the apparatus needed for ELISA is probably the simplest. That is not to say that some of the dedicated equipment is not useful and of these I would put at the top of my list an automated plate reader with computerised data processing. I have listed below what I looked for in the equipment that we bought.

Plate washer

Washing microtitre plates can be carried out with a wash bottle and a bucket as we do for some assays where a different wash solution to our normal one is used. There are, however, a number of automated washing machines on the market today which work well. Each time we review such a product we come to the conclusion that our simple set up is adequate. This consists of a reservoir containing warm solution which is placed on a convenient shelf and connected to a wash/suck manifold and a vacuum chamber.

Plate reader and data handling

Before buying a new plate reader we recently reviewed a number of different machines. All worked well mechanically but the software varied considerably with each. The software of the one we chose needed to be modified to read successive plates from a single common standard curve and has performed well since we bought it. If the machine you are evaluating can be shown to do what you want when it is demonstrated it is likely to be OK. Believe nothing unless you have seen it.

Pipettes

The same rule of try before you buy seems to apply to the purchase of pipettes too. There are a number of very accurate air and positive displacement pipettes and I always test potential purchases with a solution of radioactive protein. The type of tips used with air displacement pipettes makes as much difference as the pipette itself.

MATERIALS

Buffers

Carbonate/bicarbonate coating buffer pH 9.6, 0.1M.

Na_2CO_3
$NaHCO_3$

Take 4.24 g Na_2CO_3 and 5.04 g $NaHCO_3$ and make up to 1 litre with distilled H_2O and check pH. Store at 4°C for no more than a few weeks.

Carbonate/bicarbonate enzyme labelling buffer pH 9.6, 1M.

Na_2CO_3
$NaHCO_3$

Take 4.24 g Na_2CO_3 and 5.04 g $NaHCO_3$ and make up to 100ml with distilled H_2O and check pH. Use fresh.

Citrate/Phosphate buffer pH 5.0 , 0.2M.

$C_6H_8O_7.H_2O$
Na_2HPO_4

Make up stock solutions of 21.01 g citric acid ($C_6H_8O_7.H_2O$, 0.1M) and 28.4 g of Na_2HPO_4 (0.2M) each in 1 litre of H_2O. Add 48.5 ml of 0.1M Citric Acid to 51.5 ml of 0.2M Na_2HPO_4 producing a pH of approximately 5. Check and adjust as necessary.

Diethanolamine buffer pH 9.8, 0.05M.

$MgCl_2.5H_2O$
Diethanolamine
Conc HCl
NaN_3

Dissolve 101 mg of $MgCl_2$ in 800 ml of distilled water. When dissolved add 97 ml of diethanolamine and mix thoroughly and adjust the pH to 9.8 with concentrated HCl, make up to 1 litre with H_2O, add 200 mg of NaN_3 and store in the dark at 4°C.

Ethanolamine buffer pH 8.0, 0.1M.

> Ethanolamine
>
> HCl

Take 1.24 ml of ethanolamine solution and add 80 ml H_2O. Adjust pH to 8 with 1M HCl and make up to 100 ml with H_2O.

Phosphate Buffered Saline (PBS) pH 7.4 , 0.05M.

> Na_2HPO_4
> NaH_2PO_4
> NaCl
> NaN_3

Dissolve 16.7 g Na_2HPO_4, 5.7 g NaH_2PO_4, 85 g NaCl and 100 mg of NaN_3 in distilled H_2O, make up to 10 litres with H_2O and check pH. Store at room temperature.

Sodium bicarbonate 0.1M (approximately pH8.0)

> Na_2HCO_3

Dissolve 8.41g of Na_2HCO_3 in 1 litre of H_2O.

Sodium acetate buffer pH 5.0, 0.1 M.

> $CH_3COONa._3H_2O$
> CH_3COOH

Add 700 ml of 0.5M $CH_3COONa._3H_2O$ to 300 ml of CH_3COOH. Check pH and adjust if necessary.

Tris/HCl pH 8.0, 0.5M.

> 60.5g Tris
> (hydroxymethyl methylamine)
> 1M HCl

Dissolve the Tris in 800 ml and adjust the pH to 8.0 using 1M HCl. Make up to 1 litre and store at 4°C.

Tris/HCl pH 7.4, 0.1M.

> 12.1g Tris
> (hydroxymethyl methylamine)
> 1M HCl

Dissolve the Tris in 800 ml and adjust the pH to 8.0 using 1M HCl. Make up to 1 litre and store at 4°C.

Assay diluent

PBS or Tris/HCl /1% animal serum/ 0.5% Tween 20

> PBS, or Tris/HCl
> animal serum
> Tween 20
> Phenol red solution
> Amido black solution

To 98.5 ml of PBS, TBS or BBS add 1 ml animal serum, 0.5 ml Tween 20 and 0.5 ml of the red and blue solutions in a 100 ml bottle. Store for no more than 2-3 days. At neutral pH it should be a green colour.

Washing buffer

> PBS, or Tris/HCl
> Tween 20

Add 5 ml of Tween 20 to 10 litres of buffer (0.05%). Mount on top of a shelf or cupboard for gravity feed.

Amplified ELISA substrate diluent

> 50mM diethanolamine buffer at pH 9.5
> 1mM $MgCl_2$
> 4% v/v ethanol
> 0.1% NaN_3.

Mix reagents and check pH.

Amplified ELISA amplifier diluent

> 20mM sodium phosphate pH 7.2
> 1mM INT violet
> 0.1% NaN_3.

Mix reagents and check pH.

Additives

Animal serum

The particular animal serum used will depend on the reagents and antibodies used in the assay. Horse serum or rabbit serum are commonly used but cross reactivity between serum and reagent antibodies may make it necessary to use another serum such as goat.

Serum albumin, casein and gelatin

> 1-2% serum, albumin, gelatin or casein
> Coating buffer

All are reported to reduce background or non-specific binding in ELISA. It is generally a question of trial and error.

Coloured additives

> 40 mg Phenol red solution
> 60 mg Amido black solution

Make up phenol red (40 mg/100 ml) and amido black (60 mg/100 ml) in H_2O. Add 0.5 ml of each to 100 ml of the buffer being used.

Substrates

o-Phenylenediamine (OPD)

> 0.4mg *o*-phenylenediamine
> hydrochloride
> 0.4 µl/ml H_2O_2
> Citrate-Phosphate buffer pH 5.0.
> 2M H_2SO_4

The substrate should be made up no earlier than 10 minutes before use and hydrogen peroxide added just prior to use. Check the pH after adding the OPD as this can alter. Adjust if necessary with HCl or NaOH. Incubation time of substrate is about 20 minutes after which the reaction is stopped by the addition of 50µl of 2M H_2SO_4. Absorbance must be recorded (at 492nm) within 40 minutes since the substrate product is unstable.

p-Nitrophenyl phosphate

> p-nitrophenyl phosphate tablets
> Diethanolamine buffer pH 9.8, 0.1M.
> NaOH

Add tablets to Diethanolamine buffer and allow them to dissolve, mixing for 10 minutes. The substrate is stable at 4°C for some time. Substrate incubation time is longer than for OPD with up to 2 hours at 37°C. The reaction is stopped using 50 µl of 3M NaOH. The coloured substrate product is very stable at 4°C and plates can be stored for several days in the dark. Absorbance should be read at 405 nm.

Amplified substrate for alkaline phosphatase

NADP (tetra sodium salt)
Alcohol dehydrogenase EC 1.1.1.1.
Diaphorase EC 1.6.4.3.
Amplified ELISA substrate
Amplified ELISA substrate

The substrate NADP is purified with a suitable ion exchange resin such as DEAE or QAE Sephadex to remove contaminating NAD and made up as a 100µM solution in amplified ELISA substrate diluent. Alcohol dehydrogenase (Sigma) 70 mg + 1% BSA or similar protein stabiliser is dissolved in 7 ml amplifier diluent and dialysed extensively against amplifier diluent at 4°C. 10 mg of diaphorase (NADH dye oxidoreductase EC 1.6.4.3.) (Boehringer Mannheim) is dissolved in 5 ml of 50mM Tris HCl pH 8.0 and 1% BSA or similar protein stabiliser added. This is then dialysed against amplifier diluent. The two amplifier enzymes are diluted tenfold and mixed together immediately prior to use. The optimum ratio can be determined by mixing different ratios of them from 10:1 to 1:10 although 1:1 is generally satisfactory.

Insoluble substrate for dot blot ELISA or ELISA plaque assay

Triton X-405
MgCl$_2$. 6H$_2$O
2-amino-2methyl-1-propanol (AMP)

AMP buffer was prepared by dissolving 100µl of Triton X-405 (at 30°C), 150 mg MgCl. 6H$_2$O and 1g NaN$_3$ in 500 ml of distilled water, 98.5 ml of 2-amino-2methyl-1-propanol (AMP) at 30°C added and the mixture stirred for 1 hour and made up to 900 ml with distilled water. The pH was adjusted to 10.25 with concentrated HCl and after an overnight incubation at room temperature checked and the volume make up to 1 litre with distilled water. For the ELISA plaque assay 4ml of 5 BCIP (1mg/ml in AMP buffer) at 40°C were added to 16 ml 3% w/v agarose (in distilled water) at 60-70°C. For the dot blot assay the agarose was omitted. It is also possible to omit agarose from the BCIP in the ELISA plaque assay.

Appendix of methods
PURIFICATION OF ANTIBODIES

Affinity Purification of Antibodies

Materials

CNBr activated Sepharose 4B
Purified antigen
Antiserum
Sodium bicarbonate 0.1M pH8.0
Phosphate buffered saline (PBS)
Tris/HCl 0.1M pH8.0
Ethanolamine buffer 0.1M pH8.0
1cm x 10cm column
HCl 0.01M
NaCl 0.5M

Method

Preparation of Sepharose.

1. Add 3g (dry weight) of Sepharose 4B to 100 ml of 0.001M HCl and allow to swell up for 20 minutes. Spin down at 2000 rpm for 5 minutes and remove the supernatant.

2. Wash the Sepharose twice with PBS, spinning down the Sepharose and removing the supernatant as above. Check the pH (should be pH7-8).

3. Dissolve 1g of antigen in 10 ml of 0.1M sodium bicarbonate, pH8, measure the O.D. (280 nm) and add to the Sepharose, mix overnight at 4°C.

4. Spin down the Sepharose and remove the supernatant. Check O.D. at 280 nm and calculate the amount bound.

5. Wash the Sepharose twice with sodium bicarbonate.

6. Wash the Sepharose twice with 10ml 0.1M ethanolamine buffer, pH8.6, for 30 mins. This blocks any remaining CNBr activated sites.

7. Wash three times (5 minutes each) with 10ml of PBS.

8. Rinse once with 0.01M HCl (for 5 minutes) to remove any loosely bound protein.

9. Wash twice with PBS (5 minutes each) and check pH is back to pH7.

10. Store slurry in 0.5M NaCl at 4°C.

Purification of Antibodies

1. Take the prepared Sepharose and add up to 20ml of rabbit serum (or other serum), check the pH is between 7 and 8. Incubate overnight at 4°C.

2. Spin the Sepharose down and remove the supernatant for analysis by double gel diffusion.

3. Pour the Sepharose into a 1 x 10 cm column and wash through thoroughly with bicarbonate buffer at 30ml/hr, measuring the absorbance of the eluate at 280nm. When the absorbance is less than 0.01 continue.

4. Apply 10ml of 0.01M HCl at a flow rate of 30ml/hr, collect 2 -minute fractions into 1ml of 0.1M tris/HCl pH8. Measure the absorbance of the fractions at 280nm and determine fractions containing antibody.

5. Pool fractions required and store frozen at -20°C.

6. A successful affinity purification should yield between 6 and 18mg per 20ml of serum.

BIOTINYLATION OF ANTIBODIES

Materials

> Biotin-X-NHS
> Antibody
> Protein stabiliser (eg BSA)
> Bicarbonate coating buffer
> PBS
> Centricon™ membrane

Method

1. The antibody solution (approximately 3mg/ml) is washed twice through a Centricon™ membrane with bicarbonate coating buffer pH9.6.

2. Biotin (30mg/ml) in the same pH9.6 bicarbonate coating buffer is added to the antibody solution in a ratio of approximately 0.2mg biotin to 1 mg antibody and mix for 90 minutes.

3. Remove free biotin by washing through the Centricon membrane with PBS twice. Add 1-2% w/v protein for storage.

ENZYME LABELLING OF PROTEINS

One step glutaraldehyde method (Avrameas 1969)

Materials

> Affinity purified antibody
> 50% glutaraldehyde
> Alkaline phosphatase VII-S
> 0.5M Tris pH 8
> PBS
> Bovine Serum Albumin
> NaN₃
> Dialysis Tubing

Method

1. Two mg of antibody (2 mg/ml) and 5 mg of alkaline phosphatase (activity >1000U/ml) are dialysed together against PBS, overnight at 4°C with several changes of PBS, to remove all the ammonium sulphate present in the enzyme extract.

2. Remove from the Visking tubing and measure volume. Calculate amount of glutaraldehyde required to make a final concentration of 2% v/v glutaraldehyde. Add it to the mixture and incubate, mixing continuously, for 2 hours at 4°C.

3. Transfer mixture to dialysis tubing and dialyse against PBS overnight at 4°C with 2 changes of PBS.

4. Transfer dialysis bag to 0.5M Tris pH 8 and dialyse overnight at 4°C with 2 changes of Tris buffer.

5. Remove the mixture from the dialysis bag and dilute the final conjugate to 4 ml with Tris buffer containing 1% BSA and 0.2% NaN₃.

6. Store in the dark at 4°C.

Two step glutaraldehyde method
(Avrameas & Ternynck 1972)

Materials

> Horseradish peroxidase type IV
> PBS 0.1M, pH 6.8
> 50% glutaraldehyde
> Normal saline
> ACA 34
> Sephadex G25

Affinity purified antibody
Carbonate/bicarbonate buffer, 1 M, pH 9.5
Lysine solution, 0.2M
Saturated ammonium sulphate
Bovine Serum Albumin (BSA)
Millipore filter 0.22 μM (Flow Laboratories)
Glycerol

Method

1. Ten milligrams of Horse Radish Peroxidase is dissolved in 200 μl of PBS (0.1M,pH 6.8) containing 1.25% v/v glutaraldehyde. The solution is then left mixing overnight at room temperature.

2. The unbound glutaraldehyde is then removed either by dialysis overnight against normal saline, changing saline twice, or by passing the mixture down a Sephadex G25 or an ACA 34 column, monitoring the absorbance of fractions at 280 nm and pooling the 'activated' HRP peak.

3. The affinity purified antibody is made up to a 5 mg/ml solution with normal saline.

4. 1 ml of activated HRP is mixed with 1 ml of antibody solution. 0.1 ml of 1 M carbonate/bicarbonate buffer pH9.6 is then added and incubated with it at 4°C overnight.

5. Add 0.1 ml of 0.2M lysine and mix at room temperature for 2 hours.

6. Dialyse the mixture again PBS pH 7.2 overnight at 4°C.

7. The HRP-conjugated antibody is then precipitated by adding an equal volume of saturated ammonium sulphate solution.The precipitate is spun down and the supernatant removed.The precipitate is then washed twice using half saturated ammonium sulphate.The final precipitate is then resuspended in 1 ml of PBS.

8. Dialyse the conjugate extensively against PBS overnight at 4°C.

9. The conjugate is then spun at 10,000g for 30 minutes to remove any sediment.The supernatant is removed and BSA or HSA is added to a final concentration of 1%.

10. Filter the conjugated antibody through a 0.22 μM millipore.

11. Store the conjugate either at -20°C or, if made up to a 50% glycerol concentration using equal volumes of glycerol and conjugate, then it can be stored at 4°C.

Periodate Method (Nakane and Kawaoi 1974)

Materials

Anti-IgG immunoglobulin
Alkaline Phosphatase type V111-S P-5521
Na_2CO_3 buffer, 0.3M, pH 8.0
1-fluoro-2,4- dinitrobenzene
Ethanol
Sodium periodate, 0.08M,
Ethylene glycol, 0.16M,
Na_2CO_3 buffer, 0.01M, pH9.5
PBS
Sodium borohydride
ACA 34
Column 100 x 1.5 cm

Method

1. Dialyse 5 ml of alkaline phosphatase against 0.3M Na_2CO_3. pH8.0 to remove all $(NH_4)_2SO_4$ and stabilisers. Add 100µl of a 1% solution of 1-fluoro-2,4-dinitrobenzene, dissolved in absolute ethanol, to the dialysed suspension and mix gently for 1 hour at room temperature.

2. Then add 1ml of 0.08M sodium periodate (freshly made up) and gently mix for 30 minutes at room temperature.

3. Add 1ml of 0.16M ethylene glycol and mix gently for a further hour at room temperature.

4. Dialyse the enzyme-aldehyde solution against 0.01M Na_2CO_3, pH9.5 for 24 hours at 4°C with at least 3 changes of buffer.

5. The anti-IgG immunoglobulin is prepared by dialysing 5mg against 0.01M Na_2CO_3, pH9.5, overnight at 4°C.
6. Add the dialysed immunoglobulin to the 3ml solution of enzyme-aldehyde and mix for 2-3 hours at room temperature.

7. Five mg of NaBH is then added, dissolved, and then left to stand at 4°C for 3 hours.

8. The conjugate is then dialysed against PBS,pH7.2, at 4°C for 24 hours. Any precipitate that forms should be spun down and removed.

9. The remaining conjugate should then be applied to the 85 x 1.5 cm Sephadex G-200 column and eluted using PBS, pH7.2 with a flow rate of 5ml/hr. Collect 2ml fractions and monitor the absorbance at 403nm, the first peak should contain the conjugate.

10. Store conjugate at -20°C in 1% BSA. Do not thaw and refreeze.

ASSAYS

General comments

In the interest of simplicity, I assume the same conditions of volume and temperature for each assay. Typically these would be 100 μl and 4°C. All washes are 3X with 300μl each well. For most of my assays I use the Nunc-Immuno Plate 1 (96F) but there are many equally good plates on the market.

Indirect ELISA for IgG and IgG subclass antibodies to ovalbumin, casein or bee venom.

Materials

Ovalbumin grade VII
Casein
Bee venom
Positive serum reference pool and QC
Test sera
AP or HRP-labelled anti-mouse Ig
Monoclonal anti-IgG (8a4) and
anti - IgG subclass antibody
(NL16, Gom2, ZG4, or RJ4)
p-nitrophenyl phosphate
tablets 104-105
o-phenylenediamine hydrochloride
Assay diluent.
Wash buffer.
Bicarbonate coating buffer.
Substrate buffer.
ELISA plate.
3M NaOH
2M H_2SO_4

Method

1. Coat the plate with antigen overnight at 4°C or for 1 hour at 37°C at a concentration of between 10 and 3μg/ml in bicarbonate coating buffer.

2. Wash three times.

3. Add dilutions of reference pool (1/50-1/50,000), quality controls (1/100, 1/1000, 1/10,000) and patient sera at 1/500. Any samples outside the range of the standard curve are repeated at up to 1/50 dilution or as low as necessary. Incubate for two hours.

4. Wash three times.

5. Add anti-IgG or anti-IgG1, 2, 3, 4 at 1/1000 dilution and incubate for 1 hour.

6. Wash three times.

7. Add AP or HRP labelled anti-mouse Ig at 1/300 and incubate for one hour at 4°C.

8. Wash three times.

For Alkaline Phosphatase

9. Make up substrate in diethanolamine buffer pH9.8 as described above. Add 100µl to each well and incubate at 37°C for 1-2 hours.

10. Stop enzyme reaction by adding 50µl 3M NaOH/well.

11. Measure absorbance at 405nm.

For Peroxidase

9. Make up OPD substrate as described above, add 100µl/well and incubate at 37°C for 40 minutes.

10. Stop enzyme reaction by adding 50µl 2M H_2SO_4/well.

11. Measure absorbance at 492nm.

Read results from the reference curve generated by a series of dilutions of a positive serum pool with an IgG pan-reactive monoclonal (8a4) antibody.

Two-site ELISA for the detection of IgE

Materials

Monoclonal anti-Human IgE 7.12
IgE standards and QC
Test sera
AP or HRP-rabbit anti-Human IgE
AP-Fab' anti-human IgE
Assay diluent.
Wash buffer.
Bicarbonate coating buffer.
Substrate buffer.
ELISA plate.
Diethanolamine buffer pH9.8
p-nitrophenyl phosphate
tablets 1 04-105
o-phenylenediamine hydrochloride
3M NaOH
2M H_2SO_4

Method

1. Coat the plate with monoclonal anti-IgE at 3µg/ml overnight at 4°C or for 1 hour at 37°C in bicarbonate coating buffer.

2. Wash three times.

3. Add IgE standards, quality controls and patient sera at a 1/10 and 1/100. Any samples outside the range of the standard curve are re-assayed diluted as necessary. Incubate for 3 hours or overnight..

4. Wash three times.

5. Add AP or HRP-anti-IgE or Fab' anti-human IgE at 1/300 or 1/3000 respectively and incubate for 3 hours or overnight.

6. Wash three times.

For Alkaline Phosphatase

7. Make up substrate in diethanolamine buffer pH9.8 as described above. Add 100µl to each well and incubate at 37°C for 1-2 hours.

8. Stop enzyme reaction by adding 50µl 3M NaOH/well.

9. Measure absorbance at 405nm.

For Peroxidase

7. Make up OPD substrate as described above, add 100µl/well and incubate at 37°C for 40 minutes.
8. Stop enzyme reaction by adding 50µl 2M H_2SO_4/well.

9. Measure absorbance at 492nm.

For Ultra-sensitive assay

7. All buffers must be phosphate free.

8. 100µl of NADP is added to the wells and incubated for 10 minutes (this can be increased or decreased depending on the sensitivity required).

9. Equal volumes of the amplifier enzymes are mixed together as detailed above and 200 μl added.

10. The colour produced can be monitored and the results read at 492 nm kinetically or after stopping with 50 μl of 0.5M H_2SO_4.

Read results from the standard curve and express as IU/ml (1 IU of IgE is approximately equal to 2.4 ng).

Two site ELISA for the detection of human IgG subclass proteins

Because some of the monoclonal anti-IgG subclass antisera perform poorly when bound to plastic we use two different coating procedures depending on the level of sensitivity that we want. To obtain the maximum sensitivity the plates are pre-coated with rabbit ant-mouse IgG diluted 1/3000 in bicarbonate coating buffer.

Materials

Monoclonal anti-IgG subclass
antibody
(8a4, NL16, HP6014, ZG4, or RJ4)
IgG reference material and QC
Test sera
AP-rabbit anti-Human IgG
Rabbit anti-mouse IgG
Assay diluent.
Wash buffer.
Bicarbonate coating buffer.
Substrate buffer.
ELISA plate.
Diethanolamine buffer pH9.8
p-nitrophenyl phosphate
tablets 104-105
o-phenylenediamine hydrochloride
3M NaOH
2M H_2SO_4

Method

1. Coat the plate with anti-IgG subclass monoclonal antibody at 1/1000 overnight at 4°C or for 1 hour at 37°C.

2. Wash three times.

3. Add IgG standards calibrated against WHO 67/97, quality controls and patient sera at a 1/10,000 for IgG, IgG1 and IgG2 and 1/1000 for IgG3 and IgG4. Any samples outside the range of the standard curve are re-assayed at an appropriate dilution. Incubate for 2 hours.

4. Wash three times.

5. Add rabbit AP-anti-human IgG at 1/300 and incubate for 3 hours or overnight. It is important to add 1% normal mouse serum and 1% normal rabbit serum to the conjugate diluent one hour before needed to reduce cross reactions. The rabbit AP-anti-human IgG had been absorbed 3X against normal mouse serum agarose.

6. Wash three times.

7. Make up substrate in diethanolamine buffer pH9.8 as described above. Add 100μl to each well and incubate at 37°C for 1-2 hours.

8. Stop enzyme reaction by adding 50μl 3M NaOH/well.

9. Measure absorbance at 405nm.

10. Record the results as ng, μg or mg/ml from the standard curve. At the time of going to press there are no internationally agreed units.

Competitive ELISA for antigen

Materials

> Bee venom
> Positive serum reference pool and QC
> Test sera
> AP or HRP-labelled anti-mouse Ig
> Monoclonal anti-IgG antibody 8a4
> *p*-nitrophenyl phosphate
> tablets 104-105
> *o*-phenylenediamine hydrochloride
> Assay diluent.
> Wash buffer.
> Bicarbonate coating buffer.
> Substrate buffer.
> ELISA plate.
> 3M NaOH
> 2M H_2SO_4

Method

1. Determine the concentration of antigen that gives 50-75% maximum OD. Coat the plate with this concentration overnight at 4°C or for 1 hour at 37°C in the appropriate coating buffer.

2. Wash three times.

3. Add 50µl of dilutions (typically 200-0.02 µg/ml) of reference antigen (the same as the coating antigen).

4. Add 50 µl of the human antibody diluted at 2X the concentration that gave 50-75% of maximal binding and incubate for 1 hour.

6. Wash three times.

7. Add anti-IgG (8a4) at 1/1000 dilution and incubate for 1 hour.

8. Wash three times.

9. Add AP labelled anti-mouse Ig at 1/300 and incubate for one hour at 4°C.

10. Wash three times.

11. Make up substrate in diethanolamine buffer pH9.8 as described above. Add 100µl to each well and incubate at 37°C for 1-2 hours.

12. Stop enzyme reaction by adding 50µl 3M NaOH/well.

13. Measure absorbance at 405nm.

Plot the results as % inhibition. The % inhibition obtained with the homologous antigen (the one bound to the plate) is the reference. The % inhibition obtained with the standard tells you the amount of antigen in it. The maximum inhibition achieved indicates the degree of immunological homology.

Dot blot ELISA

Materials

>Nitrocellulose paper (HA, 0.45µM)
>Antigen
>Antibody
>Tris HCl pH 7.2
>1M KOH
>Sample
>Alkaline phosphatase labelled antibody
>Foetal calf serum
>5-bromo-4-chloro-3-indolyl-phosphate
>Triton X-405
>2-amino-2-methyl-1-propanol

Method

1. Spot 1μl of antigen in Tris HCl pH 7.2 or 1M KOH onto 1 X 10 cm nitrocellulose paper strips, dry and store at 4°C until use.

2. Block vacant binding sites with 5ml of 10% foetal calf serum or similar in Tris HCl pH 7.2 in a test tube for 1 hour.

3. Wash 3 X with ELISA wash.

4. Add 2 ml sample containing antibody in ELISA assay diluent and incubate for 1-3 hours.

5. Wash 3 X with ELISA wash.

6. Add alkaline phosphatase labelled antibody and incubate for 1-3 hours.

7. Wash 3 X with ELISA wash.

8. Add 2ml of BCIP substrate and watch blue colour develop. To stop the reaction remove the substrate and wash.

ELISA plaque assay

Materials

> Mouse anti-human IgE clone 7.12
> 5-bromo-4-chloro-3-indolyl-phosphate
> Triton X-405
> 2-amino-2-methyl-1-propanol
> Alkaline Phosphatase
> Rabbit-anti-human IgE
> Ficoll hypaque

Method

1. Coat microtitre plates with mouse monoclonal anti-human IgE clone 7.12 at 10μg/ml in coating buffer (0.1M pH 9.6 sodium bicarbonate buffer).

2. Wash 3X in ELISA wash.

3. Wash peripheral blood mononuclear cells (prepared over Ficoll hypaque) and add to the anti-IgE coated plate at $2x10^5$ per well.

4. Cell-free wells and wells to which a known concentration of human IgE had been added are included as controls.

5. Incubate overnight at 37°C.

6. Wash 3X with ELISA wash buffer which lyses the cells.

7. Add AP rabbit anti-human IgE diluted 1/300 in ELISA assay diluent and incubate overnight at 4°C or for 3hrs at room temperature.

8. Wash 3X with ELISA wash buffer.

9. Add 200 µl of substrate (BCIP 1mg/ml in AMP buffer at 40°C + 3% w/v agarose (in distilled water) at 60-70°C.

10. Within 15-20 minutes the 'plaques' began to appear as macroscopic blue spots easily visible to the naked eye.

11. The plates were left covered at 4°C overnight to allow the colour to develop and counted with the aid of a 10X magnification hand lens.

The UK suppliers we have used have included: British Drug Houses, Poole, Dorset; Seralab, Crawley, Sussex; Sigma UK, Poole, Dorset; Pharmacia, Milton Keynes.